ersitas
Mada
Jl. Solo
Jl. Laksda Adi Sucipto
To Airport
To Prambanan
To Solo
To Solo
Wahidin
Sudirohusodo
railway
Jl. Dr. Sutomo
Jl. Kusumanegara
12
Jl. Veteran
Glagahsari
Batikan
Jl. Taman Siswa
Pandean
Jl. Veteran
Bus Station
Printis
Jl. Menteri Supeno
Kemerdekaan
Pramuka
Gambiran
Ngeksiganda
Gajah Wong R.
Kemasan
Sorosutan
15
Jl. Tegal Gendu
Modorakan
N
YOGYAKARTA

YOGYAKARTA

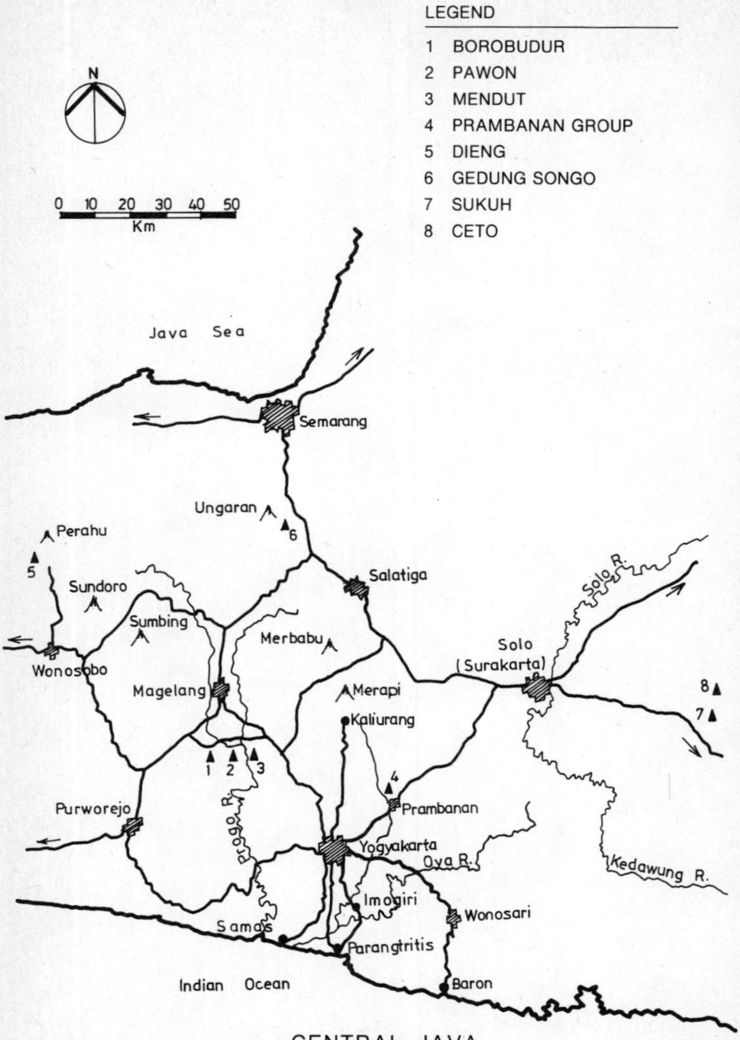


CENTRAL JAVA

# YOGYAKARTA

## Cultural Heart of Indonesia

Michael Smithies

SINGAPORE
OXFORD UNIVERSITY PRESS
OXFORD NEW YORK

*Oxford University Press*

*Oxford New York Toronto*
*Petaling Jaya Singapore Hong Kong Tokyo*
*Delhi Bombay Calcutta Madras Karachi*
*Nairobi Dar es Salaam Cape Town*
*Melbourne Auckland*

*and associates in*
*Beirut Berlin Ibadan Nicosia*

*OXFORD is a trademark of Oxford University Press*


*First published 1986*
*Second impression 1987*



*ISBN 0 19 582653 1*

*Printed in Malaysia by Peter Chong Printers Sdn. Bhd.*
*Published by Oxford University Press Pte. Ltd.,*
*Unit 221, Ubi Avenue 4, Singapore 1440*

# Preface

My first visit to Yogyakarta as a tourist fifteen years ago left a very powerful impression and I was delighted to be able to go back and live there from 1974–76, working in Gadjah Mada University. I have been back nearly every year since and the city continues to fascinate.

A guidebook to Yogyakarta was in existence ten years ago, but it is difficult to find now and has inevitably dated. This book tries to avoid the ephemeral and to be more of a companion to the visitor to the city. This has meant using a number of sources, and a particular debt of gratitude must be expressed to M. C. Ricklefs, whose two books dealing with the history of Indonesia and the foundation of Yogyakarta have been invaluable and may have been unconsciously as well as consciously quoted. Another special expression of thanks is due to Jacques Dumarçay, both for his written works, personal assistance and friendly encouragement in the area of Javanese studies over the years. My thanks also go to a number of personal friends, Yogyanese and others, for their help and support.

It is hoped that this modest handbook will help visitors appreciate one of the most interesting and relatively unspoilt cities of South-East Asia, in the same category as Luang Prabang and Mandalay, and now a good deal easier to visit than either of these.

*Singapore*
*December 1984*

MICHAEL SMITHIES

# Contents

## Maps and Plans

# Colour Plates

# Introduction

THERE is one city in Indonesia which has managed to maintain its charm and attractions, and which is relatively untouched by the modern world, and that is Yogyakarta. Its pace of life is unhurried, people remain courteous in an increasingly pressured society, and it is proudly conscious of its artistic traditions which are maintained and which still dominate the life of the city.

The city is relatively new, having been founded in 1755, but it has been since the eighth century in the foyer of Javanese culture, an incredibly refined and complex civilization relatively little known to the outside world, although it has influenced nearby civilizations and the history of Cambodia and Thailand. At present there are about one hundred million Javanese, nearly all in Central and East Java, all looking to Yogyakarta and, to a lesser extent, the other nearby court city of Solo, also known as Surakarta, for their cultural inheritance.

Yogyakarta was spelt as Jogjakarta in the early days of the Republic and Djogdjakarta under the Dutch, but still called Ngayogyakarta in Javanese. It was originally a small village and the name is apparently derived from the Sanskrit Ayodhya (Ngajodya in modern Javanese), the capital city of Rama in the *Ramayana* epic (Ayuthia, the old Thai capital from 1350 to 1767, was also so derived). The city is commonly referred to, both by its inhabitants and throughout Indonesia, as Yogya, pronounced as it used to be spelt, Jogja. Van Imhoff, the Dutch Governor-General from 1743–50, noted in 1746, before the modern city was founded, that 'it was considered the centre of Mataram', the heartland of the old Javanese kingdoms. It remains the cultural heart of the Republic of Indonesia, the fifth largest country in the world in population and also one of the largest in geographic spread. To know Yogya and its inhabitants well would give an understanding of the ways of thought and behaviour of the majority of Indonesians. But this is no easy thing; the Javanese have always been inward-looking, even before the arrival of the Europeans on the island (a glance at the map shows how the old kingdoms had their centres away from the coasts) and their civilization remains closed and difficult to penetrate.

Language is possibly a contributing factor in this respect. Javanese is a complex language with different levels. It was traditionally written in its own script, derived from ancient Indian scripts. Little writing is done in Javanese characters now and many young Javanese can no longer read it, but there is still writing in Romanized Javanese, of novels and, of course, ordinary letters. In unofficial situations, Javanese continues to be spoken by all Javanese in preference to the national language, Indonesian. By virtue of their numbers, the Javanese also exercise a strong influence on the vocabulary of spoken Indonesian.

The Javanese remain relatively untouched by contemporary materialism. They are much given to mystical interpretations of phenomena and are profoundly affected by the Hindu-Buddhist cultures which reigned supreme in the area from the eighth century up to the coming of Islam to the Yogyakarta region in the sixteenth century. They believe in the importance of assimilating into a group, of concealing their emotions and real feelings. They raise courtesy to a fine art and allow themselves the time to do so. All intrusive action and noise is regarded as *kasar*, vulgar, and the humblest and least educated appreciates the virtues of *alus*, or refined, behaviour.

The Yogyakarta region is densely populated, to the point of reaching world records in some parts. It is also, on this account, very poor, and under-employment and unemployment are major problems. For all that, it is relatively crime-free and extraordinarily relaxed and relaxing. It is also relatively cheap to visit and to live there. For these reasons, in addition to its wealth of artistic achievement and endeavour, it has become one of the locales most appreciated by the *cognoscenti* of South-East Asia.

# 1
# History

THE history of modern Yogyakarta can be dated from the Treaty of Giyanti of 1755 between Mangkubumi, the first Sultan of Yogyakarta, Paku Buwono III of Solo, and Nicolaas Hartingh, the Governor of the North-East Provinces of the Dutch East India Company, the Verenigde Oostindische Compagnie (VOC). The division of what remained of the kingdom of Mataram into two was made permanent and Yogyakarta, to which Mangkubumi had already gone in 1749, was recognized as the capital of the new principality under the Sultan. However, the selection of the site of Yogyakarta was not arbitrary.

The earliest signs of habitation are prehistoric. The skulls found in the Solo River valley of the Pleistocene Age form a link between the earliest Java Man, pithecanthropoid, and more advanced human types. Prehistoric remains have been discovered near Samas, on the coast to the south of Yogyakarta, and in other parts of the region. The whole Yogyakarta-Solo area, being extremely fertile, was an ideal site for man shifting from hunting and gathering to domestic agriculture.

The earliest archaeological remains date from the late seventh century on the Dieng Plateau. This remained a Hindu-Buddhist centre of worship at least until the thirteenth century. Hinduism and Buddhism arrived early in Java from India, some time around the first or second centuries AD, and were grafted onto the animistic beliefs of the indigenous population. The religions were almost certainly brought by traders, and rulers found in them a convenient justification and systematization of their power. The two religions coexisted in Java as they did in India, though Buddhism was never to wane so dramatically as there. In Java, there was a continuing shift between the two. The earliest temples to have survived at Dieng were Hindu, and this is true of the second oldest group, at Gedung Songo.

Borobudur was probably begun as a Hindu temple about 780 and converted to Buddhist use at the time of its second transformation about 792. This was under the influence of the Sailendra Buddhist rulers, who appear to have ruled in the northern part of the region, whereas the Hindu Sanjaya dynasty, in the southern part, had its manifestation of power and importance in the purely Hindu temple of Prambanan, consecrated in 856. The records, in the form of inscriptions, give no clear picture, but the evidence is of coexistence, if not waves of fashion. For instance, Candi Plaosan, which probably dates from between 825 and 850, and is quite near Prambanan, is a Buddhist temple with no Hindu elements, but it was clearly supported by the nobles whose names are inscribed on its stupas.

Very little is known about the origins and organization of these dynasties, though they clearly had contacts beyond Java, judging from the Javanese influence in the temples of Chaiya in southern Thailand and in some of the early Khmer temples. Still less is known about how they fell from power. One of the great mysteries of Javanese history is what caused a shift in power from Central to East Java. Wawa (?924–8) was the last Sivaite king to reign in Central Java; his successor, King Sindok (?929–47), made the move to East Java and founded a dynasty which reigned for two hundred years in the Brantas River valley. Plaosan is the last major Buddhist temple built in Central Java, and Candi Asu (between Mounts Merapi and Merbabu) and Prambanan, both dating from 856, the last major Hindu temples built in the region. No significant religious structures which have survived were built after that, though an inscription dating from 1210 has been found at Dieng, indicating that the site was still occupied at least until that date. There are also the two exceptional very late temples on Mount Lawu east of Solo, Candi Ceto and Candi Sukuh, both dating from the fifteenth century and devoted to the specifically Javanese cult of Bima, a *wayang* (shadow play) hero.

Numerous theories have been advanced for this shift: a major volcanic eruption, earthquakes, famine, war. The truth is not known. It seems unlikely that the area was entirely abandoned in favour of the Hindu-Buddhist state of Majapahit in East Java, which appears to have consolidated its power around 1294 and to have begun to collapse by the end of the fifteenth century. It could be that in Central Java more modest temples in wood continued to be built, but did not survive.

The area appears to have revived in the sixteenth century for reasons as mysterious as its previous decline. By this time, Java had become a string of Hindu-Buddhist principalities of indifferent importance, and was in increasing contact with a new religion also arriving from India, Islam, and a little later with Christianity, brought by the early Portuguese and then Dutch explorers. Islam gained ground along the coastal principalities, but Central and East Java, especially, remained Hindu. By the time of the Portuguese Tome Pires, in Malacca from 1512 to 1515, the coastal regions were Islamicized but the interior remained under the control of a Hindu-Buddhist king, resident in Kediri. However, Muslim gravestones dating from the fourteenth century have been found at Trawulan and Tralaya, indicating that Islam had reached the Majapahit courts and was at least tolerated. The Islamic coastal state of Demak finally conquered Majapahit about 1527 but its authority collapsed in the second half of the century.

In this period, two new powers arose in the districts of Mataram (around present-day Yogyakarta), and Pajang (around present-day Solo), and the power of the coastal rulers declined. Legend has it that a Kyai Gedhe Pamanahan, alleged to be (perhaps for reasons of legitimacy) the descendant of the last ruler of Majapahit, occupied about 1570 the Mataram region after a dispute. The region was said to be deserted when he occupied it, but this seems improbable. The rulers of the region appear to have been Islamicized in the first half of the sixteenth century.

His son, Senopati, enters more fully into documented history and reigned from about 1584 to 1601. Senopati, supposedly with the aid of the Goddess of the Southern Ocean, Ratu Loro Kidul, defeated the rival realm of Pajang about 1587 and took the holy regalia, the *pusaka*, the symbols and signs of power, to his court in Kota Gede. Senopati expanded the kingdom north and east and died about 1601, to be buried at Kota Gede, where his tomb remains. His son Krapyak reigned from about 1601–13 and tried to bring the powerful kingdom of Surabaya under his control. Javanese tradition has it that Krapyak built the walls of the *kraton*, or palace, at Kota Gede.

Krapyak's son, Sultan Agung (1613–46), was the most important Mataram ruler. He conquered Madura in 1624, after which he took the title 'Susuhunan', or emperor, and finally Surabaya in 1625, and the coastal regions up to Cirebon. Between 1614 and 1622 he established his court at Karta, a little to the south of Kota Gede. But the Dutch East India Company, under Coen, had established its base in Batavia in 1619 and this brought Sultan Agung into direct conflict with the Europeans, who stood in his way of uniting the whole of Java. He tried in 1628 and 1629 to besiege the Dutch in Batavia but had to retreat with considerable losses. He turned his attention to the eastern salient and in 1640 at last managed to eject the Balinese and, with them, Hinduism from Java. With this success and with the authority of Mecca, he took the title of 'Sultan', not to be outdone by the ruler of Banten in West Java who had taken the title in 1638. He then prepared his tomb on the hill at Imogiri to the south-east of Yogyakarta, and at his death in 1646 his son was declared his successor with the imperial title of Susuhunan and the name Amangkurat I (1646–77).

Amangkurat I moved the court to Plered, to the north-east of Karta and still close to Kota Gede, building in brick rather than wood. His reign was one of terror and tyranny and ended in open rebellion in which his son, Amangkurat II, was suspected of being implicated. His father having died in flight, the successor, who reigned from 1677 to 1703, took over without a court, cash or troops. He perforce had to rely on the Dutch East India Company (VOC) for support. Amangkurat I's brother, Puger, remained in Mataram and contested the succession, so Amangkurat II in 1680 moved to Pajang and established his court at Kartasura, where it remained, though frequently threatened and occasionally occupied by rebels, until being moved again to Solo (Surakarta) in 1746.

Puger remained and reunited Mataram, though acknowledged the King's authority in 1681 after failing in two attacks on Kartasura thanks to its defence by VOC troops. However, he became, with Dutch support, Susuhunan Paku Buwono I (1704–19) after Amangkurat III (1703–8), who had succeeded his father, was overthrown in the First Javanese War of Succession (1704–8) and sent in exile to Ceylon by the Dutch. The Second Javanese War of Succession (1719–23) was not long in following, the Mataram princes disputing the succession among themselves. Crises mounted at Kartasura under Paku Buwono II's inept rule (1726–49) and the King's brothers were in open rebellion: they were Pangeran Singasari, Pangeran Mangkubumi (later Sultan Hamengku Buwono I, 1749–92) and his nephew Mas Said (who became Pangeran Adipati Mangkunegoro I, 1757–95).

Prince Mangkubumi, who had returned briefly to his brother's court in 1744, accepted the Susuhunan's challenge to put down his rebel nephew holding an area to the north-east, succeeded and claimed his promised reward of 3,000 households (from which he could then draw finance and men). Mangkubumi had already objected in 1743 when his brother had ceded to the VOC the coastal area without consulting his advisors (meaning in this case himself). The VOC Governor-General appeared at court for the first time in history in 1744 and intervened in the dispute between Paku Buwono II and his brother, maintaining the gift of the households would give Mangkubumi too much power. He then capped this by publicly rebuking Mangkubumi at court for his ambition.

This drove Mangkubumi again into open rebellion and the Third Javanese War of Succession (1746–57) began. Mangkubumi joined forces with Mas Said and attacked Solo in 1748. In the midst of this rebellion, Paku Buwono II had the misfortune to die eleven days after signing a document passing over the whole kingdom to the VOC which von Hohendorff, the Governor of the North-East Provinces, had the good sense to ignore, declaring the Crown Prince the successor, reigning with the name Paku Buwono III (1749–88). Before von Hohendorff had done this, however, Mangkubumi had been declared king by his followers and assumed the title Susuhunan Paku Buwono as well. He took up residence in the region of Yogya, near the old kraton of Kota Gede, Karta and Plered.

Effectively Mangkubumi's reign dates from this year, 1749. The opposition to the declared successor in Solo multiplied and the VOC found the constant warfare, in which it was involved in its support for its chosen successor, was bankrupting it, while Mangkubumi and Mas Said appeared to be heading for a split. A Turkish intermediary was called in and negotiations were begun between the various parties (not including the declared Susuhunan in Solo) in 1754. Mangkubumi finally settled to receive one-half of the kingdom with its capital in Mataram, and half the annual rent paid by the VOC for the lease of the coastal regions, while agreeing to recognize VOC rule there. Mangkubumi agreed that two kings with the same title was inappropriate and suggested the title of Sultan of Mataram, to which the Dutch (and perforce his nephew) agreed. He took the name Hamengku Buwono I and the Treaty of Giyanti of February 1755 confirmed all these arrangements. Yogya, where he had established his court, was renamed Yogyakarta.

Mangkubumi was to prove the greatest ruler of Mataram since Sultan Agung, and for the first time since the early sixteenth century a relative period of peace was established. The first concern was the legitimization of the kingdom, through the court, pusakas, practices and marriage diplomacy. The two rulers of Central Java never met and they developed their courts separately. Law codes were agreed and relations between subjects of the two courts regulated. The division was eventually accepted as permanent; there was too much to be lost by going back. Yogyakarta became a powerful state and Hamengku Buwono I, when he died in 1792 at the age of about eighty, handed to his successor, his son the Crown Prince, a repopulated and prosperous, as well as established, kingdom.

Hamengku Buwono II, who reigned from 1792–1810, and 1811–12 and again from 1826–8, did not have his father's ability. He failed to rule by consensus and was opposed by his brothers, particularly Natakusuma; he was contemptuous of the declining VOC power but did not recognize the political changes in Europe after Holland became a vassal of France. In 1810 there was a revolt by the Sultan's brother-in-law Raden Rangga against the Europeans. The Sultan resisted the political demands of the Governor-General, Daendels marched on Yogyakarta in 1810 and forced the Sultan to step down in favour of his son, Hamengku Buwono III (1810–11, 1812–14). A large sum in prize money was taken by Daendels.

The British conquest of Java in 1811 brought new difficulties for the Yogyakarta sultanate. Hamengku Buwono II, in the confusion, took the opportunity to retake the throne. The Sultan proved no more amenable to the British than to the French-supported Dutch, and entered into secret correspondence with Paku Buwono IV of Solo (whose real aim was to bring down the Yogyakarta sultanate) to resist the Europeans. Raffles, the Lieutenant-Governor, was apprised of this and marched on Yogyakarta.

The kraton was attacked and captured in June 1812. Its contents were looted, large sums of money were taken, the Sultan was again deposed in favour of his more amenable son and Natakusuma was rewarded for his assistance with an independent inheritable domain in Yogyakarta and the title Pangeran Paku Alam I (1813–29), the precedent for which existed in Solo when the Mangkunegoro domain had been created by the Dutch in 1757. This was the only time when a Javanese court had been attacked and taken by European troops and the humiliation of the sultanate was considerable.

From 1812–25 there was increasing discontent in Central Java; Europeans interfered in court affairs, and leased lands along with the Chinese from impoverished aristocrats for plantation use, tax collectors' demands were extortionate, harvests were poor and disease spread. Hamengku Buwono IV (1814–22) was reportedly poisoned and the succession went to his three-year-old son, Hamengku Buwono V (1822–6, 1828–55). In this turmoil there arose a major revolt, the only one in nineteenth-century Java, led by Prince Diponegoro, the eldest son of Hamengku Buwono III. This started in 1825 and quickly spread, causing serious problems for the Dutch as European and Chinese lands and persons were attacked. In the midst of this the old Sultan, Hamengku Buwono II, was brought back from exile and reinstated on the throne, to no effect. The Java War lasted five years and only ended when Diponegoro agreed to enter into negotiations with the Dutch when he saw that his support was failing. Diponegoro was inevitably arrested and exiled and resistance to the Dutch ended. Thereafter, the Javanese courts were purely decorative. They concentrated on the only things left to them, the cultivation of the arts and etiquette.

Real power in Yogyakarta, as in Solo, was in the hands of the Dutch Resident and though the culture system of plantation crops officially did not operate in Yogyakarta, arrangements existed for the production of export crops. The aristocracy was tied to the Dutch bureaucracy and depended on positions in it. The Yogyakarta court re-established its library, looted by the British, and concentrated on cultural prestige. The population rose significantly.

Yogyakarta does not figure again in history until the time of the Second World War. Sultan Hamengku Buwono IX, who had been studying in Leiden, was ordered back to Yogyakarta by his father and then assumed the throne. He made attempts during the Japanese Occupation from 1942–5 at reform, but waited for the death of his *patih*, or grand vizier, in 1945 to assume control of affairs, appointing no successor. Paku Alum VIII did the same and became the Sultan's deputy in instituting reforms. The court bureaucracy was reduced, the Indonesian language officially replaced the use of Javanese, the village head tax was abolished and the electorate for village councils and headmen increased, and the Sultan created a people's army loyal to him with himself as military commander.

In January 1946, the Dutch reoccupied Jakarta and the Republic, declared on 17 August 1945 by Sukarno and Hatta, moved its capital to Yogyakarta. It remained the capital of the independent republic throughout the revolutionary period. The Sultan offered the front part of the kraton to the new Gadjah Mada University, created as all the tertiary institutions were in Dutch hands in Jakarta and Bandung. There was, therefore, no opposition from outside the palace to the palace; the revolutionaries were inside it, with the approval of its head.

Central Java was flooded with refugees, and there were armed clashes in the Yogyakarta-Solo area after the arrival of the Siliwangi division in 1948. On 18 December 1948, the Dutch launched their 'second police action', as it came to be known, and the following day Yogyakarta was taken. The republican government allowed itself to be taken, relying on world opinion to turn the Dutch victory into defeat. Sukarno, Hatta and others were captured. But the republican army could not understand this tactic and regarded the civilian politicians' manoeuvre as cowardly. It took to the countryside, killing a number of leftists in prison as it withdrew from Yogyakarta, and conducted a guerrilla war. The scene briefly shifted to the United Nations Security Council, where outrage was expressed at the Dutch action, reinforced by the fact that the United Nations Good Offices Committee trying to reconcile the two sides was only just outside Yogyakarta, at Kaliurang, at the time of the Dutch attack. The republican army declared military rule over Java and, led by Suharto, attacked and held Yogyakarta for six hours on 1 March 1949. The Dutch, under international pressure, released Sukarno and Hatta and evacuated Yogyakarta. The republican civilian government returned to the city on 6 July 1949, but the army was at first unwilling to accept its authority, and only did so when Sukarno threatened to resign. The Sultan of Yogyakarta became co-ordinator of security. Finally, on 27 December 1949, the Netherlands transferred sovereignty to Indonesia, then known as the Republic of the United States of Indonesia; this federal state did not last long and was formally replaced by the unitary Republic of Indonesia the following year. The capital shifted to Jakarta. Yogyakarta, much battered by the revolution, returned to its wonted calm.

The Sultan, however, remained involved in national destiny. He was in every cabinet from 1946 to 1953, except that of Sukiman (1951–2), and as Minister of Defence helped improve army-civilian relations. The heady years of Sukarno's guided democracy, the campaign for Irian Jaya, and confrontation with Malaysia largely bypassed Yogyakarta, though the accompanying inflation and increasing politicization of daily life did not. The events of 30 September 1965 and the attempted coup unleashed a bloodbath throughout the country, and nowhere more so than in Central

Java, before the army finally restored order. After Suharto became President, Sultan Hamengku Buwono IX served as Vice-President of the Republic for several years.

In the immediate post-war period, Yogyakarta knew an effervescence of creative as well as political activity and, as the capital of the new republic under siege, was a lively centre of painting, theatre and literature as well as the traditional arts. With the return of the capital to Jakarta, Yogyakarta lost something of this impetus, though it remains the undisputed traditional cultural capital.

# 2
# The Palace Complex

## The Kraton

THE kraton, or Sultan's palace, of Yogyakarta was founded by the first Sultan, Mangkubumi (Hamengku Buwono I), in 1756 and the original buildings were said to be completed a year later. It is possible for most of the buildings to have been finished fairly quickly, since the greater part consist of open Javanese pavilions known as *pendopo*, and the outer fortifications were not constructed until later in the reign. Prior to taking residence in the kraton, Mangkubumi inhabited a temporary residence, Kraton Ambar Ketawang, quite close to Yogyakarta, a few kilometres to the west near the village of Gamping. The present kraton consists of a series of interconnecting rectangular open courtyards, with open pavilions of various sizes in each, the whole surrounded by high whitewashed walls.

The kraton is effectively a town within a city, since behind the high, thick, brick outer walls, each a kilometre in length, a huge number of people live and work. The walls were not meant to be merely decorative. Shortly after he assumed the Lieutenancy-General of Java in 1811, Raffles had trouble in assuring the submission of the 'Court of Djocjocarta' and went, in June, with Major-General Gillespie in charge of a small body of some 1,200 European troops and Indian sepoys, supported by 800 men of the Mangkunegoro Legion, to force the second Sultan to terms. Raffles wrote to Minto, the Governor-General in Calcutta:

> I should mention that the Craton was a regular fortified position about three miles in circumference, surrounded by a wide and deep ditch, with a wall forty-five feet high, defended by well-constructed bastions, and forming ramparts all round.
>
> The approach to the Craton being further secured by lower walls within the ditch on the opposite side of the road, and the gates protected by drawbridges after the European model; at the period of the assault it was

calculated that there could not be less than 11,000 armed men within the Craton....

The assault was made by escalade; we soon got possession of the ramparts, and turned their guns upon them ... at nine o'clock the Craton was ours ... the Sultan was taken in his strongest hold....

He was deposed by Raffles for his opposition, to be replaced by the Crown Prince. Hamengku Buwono II was exiled to Penang, the kraton plundered, his brother, Natakusuma, who had given Raffles assistance, was given the title Paku Alam I and lands; the Yogyakarta principality was split into a senior and junior section.

John Crawfurd had been sent by Raffles to Yogyakarta in 1811 to be the Resident, and his inflexibility was to contribute to some degree to the ultimate attack on the kraton. He was later to be a governor of Singapore. He gave a description of the Javanese kraton in his *Descriptive Dictionary of the Indian Islands and Adjacent Countries*:

The palace of the prince is called *kraton*, a derivative from *ratu*, a king, and meaning the royal residence. This may be considered a walled town. The actual palace occupies the centre, and is surrounded by the dwellings of the princes, and those of attendants and retainers. The spaces unoccupied by houses contain the gardens and reservoirs of the sovereign. The principal approach to the palace is from the north, and through a square or court of considerable extent called the *alun-alun*, the sides of which are adorned by rows of fig trees (Ficus benjamina), while a pair of these are invariably in its centre. It is here that the prince shows himself to his subjects with much ceremony once in every week, and where tournaments, public processions, and military exercises are exhibited.

The external walls of the ancient *kraton* were of hewn stone, or of excellent brick and mortar, without any other defence than round towers.... Of the extent of these walled towns we can judge from the modern one at Yogyakarta, which is three miles in circumference, and in my time contained a population of 10,000 inhabitants....

The construction of these walls caused some misgivings to the Dutch. In a fury of activity, in complete contrast with the continual postponement of the Sultan's promised fort to accommodate the Dutch, a massive wall was put up in two weeks around the kraton in 1785 at the instigation of the Crown Prince. The architect was

Tumenggung Wiraguna, the Crown Prince's patih, or grand vizier, who had spent two years at Batavia in charge of the Crown Prince's troops, and the walls were said to resemble closely the Dutch fortifications there. The Dutch were relieved to learn that the Sultan's cannon could not fire on the incomplete Dutch fortress, but nevertheless had the fortifications measured by people from the Semarang Marine Academy.

The main entrance to the kraton is reached off the Alun-Alun Lor, the open space to the north of the kraton area, which Yogyakarta's main street, Malioboro, leads into (Plate 1). Facing the Alun-Alun Lor are a number of open pavilions which are not normally visited. These are used during special ceremonies, usually accompanied by gamelan playing and court dancing. It was here that Indonesia's first university, Gadjah Mada, functioned until it could construct its own buildings.

To reach the public entrance to the kraton, one should go through the Alun-Alun Lor to the south, to the formal entrance and pavilions just described, and walk around the walls to the right (that is, keeping one's left side by the kraton wall). In this way one reaches the outer entrance courtyard, which lies behind the buildings facing the Alun-Alun.

From the outermost courtyard, Keben, one passes through the Srimanganti gate to enter another courtyard, with a pendopo on either side. To the right are housed two gamelan orchestras, Nagawilaga, tuned to the scale of *slendro* (five almost equal distances per octave), and said to be two hundred years old, and that of Gunturmadu, tuned to the scale of *pelog* (seven unequal distances per octave), said to be five hundred years old. These are taken to the Grand Mosque, off the Alun-Alun, once a year and played for the Prophet Mohamed's birthday. On the left of this courtyard is a pendopo containing palanquins for weddings, litters, chairs for the application of cosmetics and a cage in which the Sultan's children were placed when they were allowed to touch the ground for the first time at the age of seven months. Inside the cage were placed various toys and objects and, depending on which the child picked up first, one was supposed to be able to determine his future character.

At the far end, to the south, are two rather ugly silver-painted *raksasa*, or giants, obviously copies of those to be found guarding the entrances of the temples of Candi Sewu and Candi Plaosan near Prambanan. That to the left of the doorway is said to have mainly good characteristics and is named Cinkorobolo, and that to the right, opposite, is called Bolouboto, the reverse in character. Beyond them, suspended from the ceiling of the Donopertopo gateway, are two good crystal chandeliers. In all, there are said to be nine gateways in the kraton proper, symbolizing the nine orifices of the human body, though the Javanese love of symbolism does not appear to be pushed to the extent of indicating which gateway represents which corporeal opening.

This entrance leads into the central courtyard, Plateran Kedaton (Plate 2). Immediately to the right is the Sultan's private office, Gedung Purworetno (Plate 3), and beyond it the private residence of the Sultan, called Gedung Kuning, from the three yellow doors on the outside. The pusaka, or royal heirlooms, are kept here and comprise ceremonial swords known as *kris*, manuscripts, musical instruments and other precious objects. The gallery outside the building contains three good nineteenth-century European mirrors, side-tables and chandeliers and is decorated with Chinese ceramic stands topped with ornamental flower pots.

The central pavilion, Bangsal Kencono, or Golden Pavilion, has four main teak pillars representing the four elements, and there is a wealth of other pillars around with elaborately decorated bases and capitals. There is a strong element of colour symbolism here, with a red motif, said to be Hindu in origin, at the top of the columns, a golden design based on the Buddhist lotus leaf, and below in black and gold the opening words of the Koran, 'There is only one God' in a vaguely Cufic character. The roof sweeps down from its central point, covering the marble floor.

Beyond this pavilion, to the west, is the Bangsal Proboyekso and the Gedung Keputrian, the residence of the Sultan's family, and to the south is the Bangsal Manis, where food and drink were served. This was restored in the Javanese year 1853 (AD 1925) as the symbolic devices, known as a chronogram, indicate in the head of the giant Condrosengkolo. The building is not very lovely with its rusting corrugated iron canopies and incongruous stained glass clerestories.

The central courtyard contains three small pavilions facing the Golden Pavilion, where court musicians play and attendants wait during ceremonies. It is planted with *sawo* trees and edged with buildings serving different purposes. To the south, leading to private apartments, the Kemagangan gate has the symbolic double naga snake with the name Dwinogorosotunggal, giving, in reverse, the Javanese calendar date of 1682, or AD 1756, the date of the foundation of the kraton. To the left of this, in the south-east corner, is a tea-house which doubles as the Sultan's bank, and on the lower east side of the courtyard is the Gedung Sarangboyo, where alcohol was served to Dutch and other infidel visitors.

To the east there is an entrance to another courtyard, Kesatrian, with interesting cabinets on either side beyond the massive gateway. Beyond this, again with two Indo-Dutch cabinets on the inside, is a further courtyard with, in the central pendopo, a gamelan *komplet* (that is combining both slendro and pelog tunings), which is played every Monday and Wednesday, and on Sunday morning with court dancing.

There is a collection of royal portraits in a pavilion enclosed on three sides at the eastern end of this courtyard, in an indifferent state of preservation perhaps reflecting their artistic merit. Some, however, are by Raden Saleh (?1814–80), who spent over twenty years in Europe and was a court painter of some renown. At the southern part of the courtyard is a small collection of pictures and photographs relating to previous sultans and grand viziers.

To the north of the outer courtyard of Kesatrian is the palace museum, Gedung Kopo, comprising mostly presents to various sultans, in the form of crystal lamps, medals, and Chinese and Japanese porcelain. To the left is the Sultan's throne, with royal betel-nut tray to one side and spitoon to the other, as well as the regalia, collectively known as the *upacara dalem*: a gilded peacock (for dignity), a cock (bravery), a deer (speed of mind and action), a goose (purity and watchfulness), a king snake (power and responsibility), a golden handkerchief box (cleansing the soul of stains), a powderbox (benevolence), and a lantern (illuminating the people). These are, in fact, copies of the real impedimenta kept in the Sultan's private quarters.

With the exception of the main pavilion in the central courtyard, the kraton is often architecturally undistinguished and sometimes of doubtful artistry. It is, however, redolent of atmosphere to a tremendous degree. The aged male palace retainers, many of whom are official guides, are dressed in traditional court attire of *blangkon* (head-dress), *surjan* (jacket), cummerbund with kris tucked in the back, batik sarong and bare feet. Their female counterparts wear a black *kebaya*, or blouse, batik sarong, carry enormous handbags with an umbrella sticking out and huge snoods of usually false hair covering the nape of their necks. Scantily-clad workmen go about their business of repair and maintenance, sweeping and arranging. All contribute to the powerful feeling of lost time. The kraton does not set out to impress, as a Western palace does, in height or luxury. Yogya's frequent earthquakes do not permit height, and furthermore, there is no tradition, at least until recently, of building residences in stone or brick, materials reserved respectively for temples and fortifications. The kraton is worth visiting for the impression it gives of being set apart from time and from the twentieth century, and in this it can be said to be typical of Yogyakarta as a whole.

On the eastern side of the kraton, inside its outer walls, are several quarters originally inhabited by people with palace connections. This sector is entered through the massive gateway of Plengkung Wijilan, under which all traffic must pass. Though now somewhat crowded and occasionally tawdry, a few elegant homes remain and gamelan playing and classical dancing are much practised here. No foreigners are allowed to inhabit the area within the outer kraton walls.

To the south of the kraton, and still within its outer walls, is another open space, the Alun-Alun Kidul, or southern square, the counterpart of that to the north, but smaller, though like it also decorated with two spreading *wringin*, or banyan trees. Herds of goats are frequently found grazing here. The central building facing the Alun-Alun Kidul, Sasono Hinggil, is used for infrequent shadow play performances of reputed quality. Leading out from this is the southern fortified gate of Plengkung Gading (Plate 4); as for the Wijilan gate to the north-east, all modern traffic has to pass under this.

## Ngasem

Ngasem, the bird market, is located quite near the kraton, to the south-west. To reach this from the kraton, one walks down the road known as Rotowijayan; one building in this street, from which it gets its name, houses the ceremonial carriages of the Sultan. The building is not normally open to the public. Turning left into Ngasem, one comes to the market area. Ngasem is an ordinary daily market, with the usual interest and squalor combined, but in the inner part it specializes in the sale of birds of various kinds, mostly those found in Java, but also some from the outer islands. All around the market in the street are stalls and shops selling seed, millet, batik dyes and tracers, and birdcages of various shapes and sizes. These last are incredibly cheap and make attractive if bulky souvenirs; with paper stuck on the insides, they also make unusual and inexpensive table or ceiling lamps which are a good deal more aesthetic than those to be found in the shops.

## Taman Sari

Taman Sari, called the Water Castle in English from the Dutch Waterkasteel, actually translates into English as 'Fragrant Garden'. It can be reached on foot from the south of Ngasem, which in fact occupies part of the huge artificial lake that once surrounded the large building to the south of Ngasem, called Kenongo. Taman Sari presents today an interesting if chaotic ensemble of ruins, Javanese kampong houses (mostly given over to selling batik pictures to tourists), watercourses and vegetation. An accurate plan of the original state has been given by Dumarçay (*BEFEO*, Vol. LXV, 1978, p. 598) but this does not take account of its condition today, which is radically different. The whole area is labyrinthine. Not to be missed, and assuming they can be found, are the sunken circular building, the restored sunken passage with pergola openings at regular intervals, and the vast ruined hall of Kenongo. The restored sunken bathing pools are of lesser interest; an entrance fee is charged for these. Without a good sense of direction, it is best to take a guide. There is no shortage of young men in the area who will probably want to show the batiks they are selling as well. The batik paintings to be found on sale here are remarkable more for their cheapness than for their originality but make decorative presents.

Taman Sari was built by the first Sultan of Yogyakarta, Mangkubumi, and started almost at the same time as the kraton in the middle of the eighteenth century, after the founding of the principality. The *Babad Mangkubumi* gives the date its construction started as 1683 in the Javanese calendar, or AD 1757–8. The completion of the whole must have taken many years. It was a pleasure garden on which no expense was spared. Dutch visitors in the late eighteenth century described in amazement its underground passages, sunken basins and refinement. It was built with much more solid materials than the kraton itself, though the kraton had stout walls (described by Raffles in his attack in 1812), most of which remain today. Taman Sari was constructed with bricks, a readily obtainable material from local clay, and several buildings, notably Konongo, that in ruins immediately to the south of Ngasem, were two storeys high, a rarity in Yogya, and higher than anything inside the kraton. The bricks were covered in three layers of stucco. The architect was a Javanese from the court and the techniques used, notably in the arches, were rather primitive, so that when it first fell out of favour, as seems to have been the case after the death of its creator, Mangkubumi, in 1792, it was not slow to fall into ruin.

Apart from its purely decorative aspects, it may also have been conceived as a retreat in times of war (though it was not used as such in 1812). The Dutch from 1765 were continually trying to get the Sultan to complete, as agreed, their fort, but the Javanese, unsurpassed masters at rationalized procrastination, promised more and did less, while at the same time spending considerable energy and sums in completing the Waterkasteel and other projects. Van der Burgh, the VOC Governor of the North-East Coast at Semarang, complained in 1775 that the Sultan 'occupies and exhausts his subjects incessantly with building all sort of... constructions for his pleasure, and demolishing and reconstructing them if they should not satisfy intentions or expectations'.

The original Taman Sari was in three parts. The upper part, to the north, consisted of the large pool, Segaran, with three structures rising from its waters, the largest of which was Kenongo. The middle part to the south, contained gardens of coconuts, cloves, pepper and the Umbul Binangun, 'the fountain of youth'. The lowest part, to the south-west, contained vegetable gardens, mangoes, pineapples and the Gedung Blawong, said to be the Sultan's bedroom, and other buildings for the kitchens and gardeners.

The island structure of Kenongo consisted of three terraces, with the two-storey building surrounded by four pavilions. There were two staircases leading to the upper storey which were partially built on the outside. The lower floor was vaulted. This was the largest and most imposing building in the whole of Taman Sari, a visual focal point to which visitors, like the Governor of Semarang, Jan Greeve, in 1788, were taken by gilded boat across the huge pool, more like a lake, of Segaran, and were duly impressed.

Kenongo was reached either from one of three projecting jetties on the edge of the pool Segaran, or by taking one of two underwater passageways, each linking one of the two other buildings projecting above the waters. The passage from the south had five lanthorns, or pergolas, rising above the level of the water in the pool and gave light and air to the passageway (which has been restored). It also gave access to Pulo Panenbung, a rectangular building on three levels, one below and two above the water-level, which may have served as a dining-room, since one of its names is Rumah Makan; it is also known as Gedung Cemeti, the 'whip pavilion' (Plate 5).

The underwater passage from Kenongo to the west is today mostly blocked, but linked the central island with the most curious building in the complex, the Sumur Gemuling (a well in the

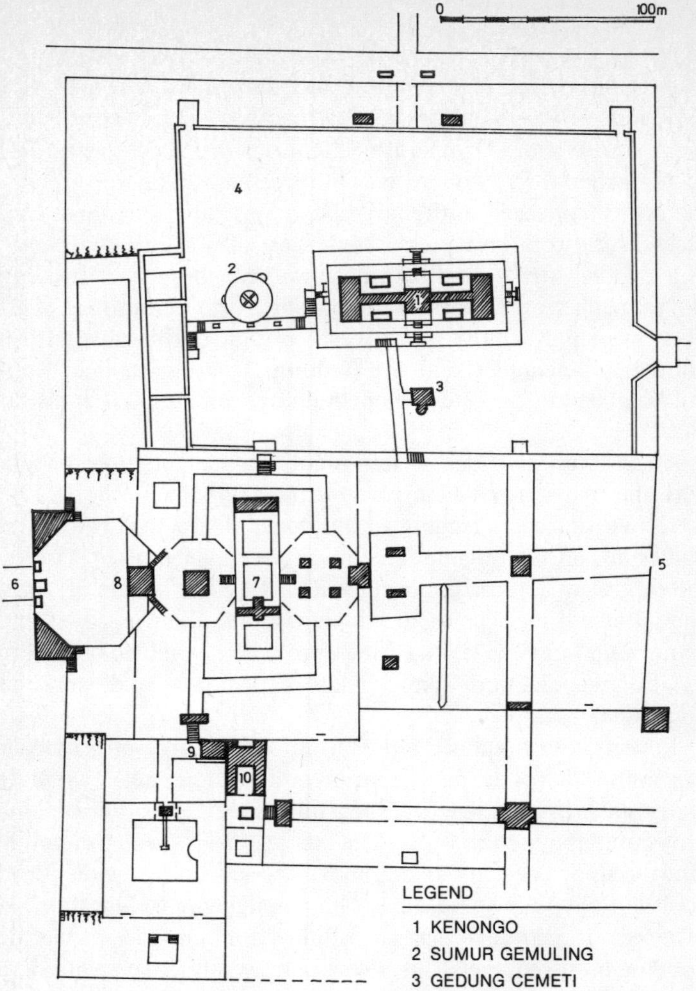


TAMAN SARI
ORIGINAL PLAN

LEGEND

1 KENONGO
2 SUMUR GEMULING
3 GEDUNG CEMETI
4 SEGARAN
5 PRESENT DAY ENTRANCE
6 ORIGINAL ENTRANCE
7 UMBUL BINANGUN
8 GEDUNG GAPURA AGUNG
9 GEDUNG DAPUR
10 KEDATON PASAREYAN DALEM
(the Sultan's Bedchamber)

'Water Palace at Joejo Carta' from W. Thorn, *Memoir of the Conquest of Java* (London, 1815).

shape of a bolster or similarly round object (Plate 6)). This circular construction comprised two levels, one above and one below the water level. The lower is like a ring built around a wide central well, for the most part filled in today, but which had a water level lower than that of the pool Segaran outside. There are two projections from the circle, an alcove to the west which apparently served as a mihrab, facing Mecca, and a slightly bigger one to the north-east which housed the latrines. Four narrow brick stairways reach out from the lower level over the well, to meet at a small landing in the middle and from there a single stairway goes to the east to reach the upper level. The upper level also has the same projections as that below it, but was also lit by windows slightly above the level of the waters outside. The building would appear to have served some religious purpose, not only in regards to Islam, but also including Javanese Hindu-Buddhist ideas. The Sultan was explicitly identified with the god Vishnu, whose statue was traditionally placed above a well containing water.

The original entrance to Taman Sari was from the west into the middle part. Most visitors now approach the complex from the road from the east, leading down a broad path with kampong houses on both sides to the restored Umbul Binangun. This has a rectangular central part comprising three pools and two buildings at a lower level than the octagonal courtyards to the east and west of the pools. The northern building contains on its eastern side a masonry couch. To the west of the western octagonal courtyard is another, larger octagon, on the western side of which is the most elaborately decorated building of any in Taman Sari. It is known as Gedung Gapura Agung. Its restored east wall, similar to that on the outer side of the eastern octagonal courtyard, is in relief, showing largely vegetal decoration.

The lowest level, to the south-west, leads to the remains of the buildings known as Gedung Gerji (tailor's building), Gedung Dapur (kitchen), Gedung Lawon (cloth store) and the Kedaton Pasareyan Dalem (the Sultan's bedroom), a U-shaped building, in the central part of which are two couches or beds which may have appeared to float on water. Water flowed through the north of the building and took refuse away from the latrines. Below this was another pool with a belevere and the lowest pool of all was that set aside as the bathing place of the gardeners.

The whole of Taman Sari depended on very carefully planned irrigation, starting on the north-east and flowing through to the south-west, through a series of pools, sluices, and controls. Much of the system was underground, in vaulted brick watercourses, representing considerable ingenuity and investment.

It is not certain when the palace fell into disrepair, though its invasion by squatters is relatively recent. Most probably it was after the death of the first Sultan, and this may be why it was not used as a retreat in 1812, at which date it was already noted by the British, who were fascinated by the complex, as being neglected. Any tropical garden and building needs constant upkeep, and the smallest negligence is rewarded with swift decay. Yogya suffered a particularly devastating earthquake in 1867, which must have contributed to the ruinous state of Taman Sari.

However, changes were made to the canalization in the nineteenth century, and these may well have been part of the restoration of the kraton undertaken from 1814. It seems, therefore, that Taman Sari was in part abandoned, restored and slightly altered (some pools and canals were filled in, others constructed) and then abandoned again.

Eliza Scidmore, the American traveller, described it in 1899 as:

> ... an Oriental Trianon, a paradise garden of the tropics, where former greatness spent its hours of ease in cool, half-underground chambers and galleries.... The Taman Sarie is sadly deserted now.... The ornamental waters are choked with weeds and rubbish; the carved stonework is black with mould and lichens; the caves, grottoes, tunnels, staircases, and galleries around the wells are dripping and slippery with green mosses; and the rose-gardens and shrubberies are fast going to jungle.

A restoration project was established a few years back but, perhaps fortunately, not much came of it. The sunken bathing pools and the associated buildings of Umbul Binangun to the east and west were restored (in grey cement) as well as the decorative wall of Gedung Gapura Agung to the west, and the pergolas and passageway of the southern sunken approach to Kenongo. There the project seems to have stopped. Swift climatic weathering has given a patina of age to these somewhat infelicitous renovations. It would be virtually impossible to restore Taman Sari to its original state, for it would involve the wholesale removal of all the numerous occupants of the kampong houses, who now presumably have some tenancy rights, and even if possible is probably not desirable: its artistic value would be marginal and its upkeep phenomenal. One will have to be content with it as it is, an attractive, fascinating muddle. There is, in any case, an attraction in ruins which complete (and restored) buildings often lack.

A host of tales abounds about sunken passageways connecting Taman Sari with the kraton proper (not impossible, but none has been discovered) and with the sea (entirely improbable, given the distance). Certainly the place abounds with symbolism in the best Javanese tradition. It probably represented for Mangkubumi the legitimization of his reign, combining in the Javanese syncretic tradition Islam, Hindu-Buddhism and establishing an elemental link with the Goddess of the Southern Ocean, Ratu Loro Kidul, to whom his descendants' hair and nail parings are offered on the shores of Parangtritis. The *Serat Surja Radja*, a manuscript kept in the kraton, specifically states that the Goddess of the Southern

Ocean retires to meditate to a mosque in Taman Sari, and in the *Babad Kraton*, Sultan Agung, the greatest of the Mataram rulers, visits the Goddess in Taman Sari. Mangkubumi was establishing his link both with the Goddess and with his most illustrious predecessor in creating Taman Sari.

But this pleasure garden was also in the oriental tradition of such constructions, whereby the heat is allayed by flowing water and luxuriant natural growth, and the Sultan could dally with his favourite consort on water-cooled couches or bathe in sunken pools hidden from the public eye. It was meant to be a sensual delight, and the vegetation was not merely utilitarian, but designed to produce particularly attractive odours from fruit trees and vines at chosen spots in the complex.

## Krapyak

To the south of the kraton complex and the Alun-Alun Kidul, at the end of the road called Jalan May. Gen. Panjaitan leading from Plengkung Gading, a couple of kilometres from the kraton, is the building known as Adukidang (Plate 7), in the village of Krapyak. *Adu* means fight or contest, and *kidang* in Javanese is a deer; *krapyak* is a fenced-in game reserve. While it is sometimes held to be a fort for soldiers (*prajurit*) belonging to the kraton, its construction, as well as its name, lead one to believe it was more likely an elaborate rest-house or hunting-lodge. It is a massive structure of brick, similar in many ways to the building Kenongo in Taman Sari, with a vaulted central area opening to four sides. The upper floor appears to have been reached not by a staircase but by a wooden ladder.

Mangkubumi, the first Sultan of Yogyakarta, retired to Krapyak in 1791, in the last year of his long life (he died in 1792 when he was about eighty years old), possibly bored with the luxuries of both the kraton and Taman Sari. It may have been to this building that he repaired.

Further along the road beyond Adukidang, past fields of sugarcane probably introduced in the nineteenth century under the Culture System, is a village where gamelan instruments are still made.

# 3
# The Town

## Alun-Alun Lor

THE Alun-Alun Lor has already been mentioned in relation to the kraton, in front of which it was the main meeting-place of the city. Major spectacles used to be and still are held here. From the highest part of the kraton, the raised terrace to the north called Siti Inggil Ler (Tradag Sitinggil), probably built in 1769, the Sultan and his guests used to watch the fights organized between tigers and buffaloes which became towards the end of the reign of Mangkubumi, the first Sultan, a regular show put on for each visiting Dutch governor from Semarang.

The Dutch party did not apparently realize that for the Javanese the contest was highly symbolic. Raffles seems to have been the first Westerner to have noted the symbolism:

> A favourite and national spectacle is the combat between the buffalo and the tiger.... It seldom fails that the buffalo is triumphant, and one buffalo has been known to destroy several full grown tigers in succession.... In these entertainments the Javans are accustomed to compare the buffalo to the Javan and the tiger to the European, and it may be readily imagined with what eagerness they look to the success of the former. The combat generally lasts from twenty minutes to half an hour, when, if neither of them is destroyed, the animals are changed....
>
> A hollow square of spearmen, four deep, is formed on the *alun-alun*, in the centre of which are placed the tigers in small separate cages.... Two or three men, accustomed to the practice, at the command of the sovereign, proceed into the centre of the square, ... and drawing the wooden door up, throwing it to one side, themselves retreat from the spot at a slow pace, to the sound of music....

The tiger represented the foreigner, more particularly the Dutch, ferocious, rapid to attack, but quick to give up and retreat. The buffalo, standing for the Javanese, was restrained in movement but with a hidden inner strength which in the end overcame his apparently stronger adversary. The applause which greeted the almost invariable defeat of the tiger was always marked.

The *gunungan*, the mountain-shaped figure symbolizing great natural forces which is shown in the *kayon*, the marker of the beginning, end and changes of scene in *wayang kulit* (shadow play) performances, often includes representations of a buffalo confronting a tiger, and appears to pre-date the arrival of the Europeans. The buffalo also became the symbol of the Indonesian Nationalist Party (PNI) in the present century. The buffalo is therefore a very ancient symbol in Java, and of course remains a common domestic animal essential for rice growing. Tigers used to be far from rare in the Javanese uplands. Even today, when Mount Merapi is about to erupt, tigers come down from its slopes and sometimes enter the city, where they are caught as soon as possible and sent off to the zoo. They are fed, incidentally, on stray dogs which the municipality collects.

When Greeve, the Dutch Governor of Semarang, paid his first and not entirely successful visit to the court of Mangkubumi in August 1788, he was treated to a tiger-buffalo fight in the Alun-Alun. The tiger soon became bored and retreated to a corner; his boredom was equalled by the Governor's, who suggested that a second tiger be introduced into the fight, to join the first against the buffalo. The idea pleased the Sultan who was interested to see what would happen. The second tiger at first fought furiously, but the first tiger did not join him, and the second soon also retired to a corner, where it was joined by the first tiger. Both tigers were then prodded with burning torches to encourage them to resume the fight, but instead of attacking the buffalo they fought each other, and in a short while both died, probably more from the mutually inflicted wounds than from those of the buffalo. The innovation pleased the Sultan as much as the Governor, who undoubtedly did not see the significance of the outcome. Even when the tiger had reinforcements, the buffalo won, and the foreigners ended by fighting each other, leaving the field empty and the Javanese victorious. Sultan Mangkubumi said that he intended to repeat this innovation.

A further fight was put on in September 1791, towards the end of Hamengku Buwono I's life, for visiting commissioners for military affairs sent out to the Indies by the Netherlands government. The commissioners, according to a contemporary chronicle, were supposed to have presented to the Sultan a particularly strong tiger, which had the added distinction of being white. Needless to say, the buffalo defeated the tiger, to the delight of the Sultan, who gave the body of the dead tiger to the Dutch Resident. The unusual colouring of the animal may more likely have been poetic licence to reinforce the tiger's symbolism.

The advanced hydrological techniques displayed in Taman Sari were also utilized in relation to the Alun-Alun, which could be flooded to put down the dust caused by dry weather or tiger combats.

Today the annual Sekaten fair, which lasts a week and occurs in the week before the Muslim feast of Maulud, the anniversary of the Prophet Mohamed's birth, certainly fills the square with dust and noise, as well as visitors from outer villages and, it is said, the occasional pickpocket. Other shows are sometimes held here, but, when the square is not occupied in this way, it acts as a sports ground for those schools in the vicinity which do not have their own playing-fields.

The Grand Mosque (Masjid Ageng) is located on the western side of the square. It was built early during the reign of Mangkubumi, in the Javanese year 1699 (AD 1773) according to one text, and is in the form of a traditional Javanese pendopo. It scarcely looks like a mosque and seems very far from grand with its rusting corrugated iron roofing. The two state gamelans are carried here in procession during the Sekaten week, leading to the Garebeg Mulud festival, and played continuously for several days in a rather mournful fashion; the tunes played are said to be extremely ancient.

Still in the western half of the Alun-Alun, but on the northern side, is the Sono Budoyo Museum. This was commissioned in 1933 by Dr Bosch, the Director of the Java Institute, a Dutch organization set up to pursue Javanese studies. The Dutchman Thomas Karsten was appointed architect and the buildings were meant to be temporary; half a century later they still stand and serve. They follow the general plan of a kraton, laid out around a pendopo, backed by a passage walkway, or *pringgitan*, leading to the *dalem*, or inner royal quarters. The museum originally included an art school, now abandoned, and the present library was added later. The director's residence, located in one corner on the first floor, is original. The buildings were opened in 1935 by Sultan Hamengku Buwono VIII, with Bosch and Karsten speakers at the ceremony.

The collection consists of arts and crafts from Java, Madura and Bali. There are some excellent artifacts from early Javanese civilizations, especially of the eighth–ninth centuries. There are also two ancient gamelans, many wayang kulit and other puppets, and numerous examples of wood carving; the collection of Balinese carvings is suprisingly good. As with most things in Yogya, the museum is open only in the mornings.

### Malioboro

Leaving the Alun-Alun Lor and going north, in a straight line from the kraton, one comes to a road junction. The main post office, Kantor Pos Besar, is to the right, a fine period piece with art nouveau stained glass and marble counters. Straight ahead is the main shopping street of Yogya, Malioboro, the first section of which is actually called Jalan Jenderal A Yani.

It is the Westerners' legend that the name Malioboro is derived from the Duke of Marlborough, but this would seem to be folk etymology. There is no reason why the Javanese should name the most important street in the city after an English general who died in disfavour more than thirty years before Yogyakarta was founded. Nor is it likely that they would name the street from a popular French soldiers' song 'Malbrough s'en va t'en guerre'. The name appears to be derived from Sanskrit terms meaning 'garland-bearing street'.

At the bottom end of the street which is in effect Malioboro, on the right, opposite the main post office, is the fort Benteng Vredeburg, which has been converted to a museum showing diorama of the Indonesian revolution. The restoration has been somewhat heavy. This fort, with solid whitewashed brick walls, was the original Dutch fort in Yogya and its name is almost certainly a corruption of the name of the Dutch Governor of the North-East Provinces, J. R. van der Burgh (1772–80), who tried so hard to get Sultan Mangkubumi to complete the promised fortifications for the Dutch (the Sultan preferred, as has been seen, to exert his building interests in Taman Sari and his own fortification of the kraton).

Early in 1765 the Sultan had at last agreed to a frequently-made Dutch request for the construction in Yogyakarta of a fort to house the VOC's troops. The Governor at Semarang, van Ossenberch, foresaw rapid progress. His successor, Vos, who took over later in the same year, anticipated, in a letter to Batavia, the completion of the fort by the end of 1766. His successor, van der Burgh, reported that the Sultan was happy to build the fortress at his expense but had to write in a letter to Batavia in 1779 that, after fourteen years of construction, the fort 'has made little progress' and told his successor in 1780 that the reason for this was the construction of the pleasure buildings of the Sultan and the Crown Prince. In 1781, Siberg, van der Burgh's successor, after a visit to Yogya, sent Batavia an encouraging report on the progress of the construction of the fort, but in 1787 was in turn to tell his successor, Greeve, that he hoped the building would be finished in two years. In 1790 the work was said to be nearly finished, but was still not complete.

This tale of deliberate procrastination was in complete contrast with the speed with which the outer walls of the kraton had been erected in 1785 and which had caused many misgivings to the Dutch at the time. The Dutch government commissioners for whom the tiger-buffalo fight had been arranged in 1791 reported on the strength of the kraton walls and noted that little could be done against them by the Company's fort. The fort was, however, able to accommodate Raffles' troops in 1812 prior to the storming of the kraton, and is shown on a contemporary map of the city.

Opposite the Vredeburg fort is the Gedung Negara, or state guest-house. This was the original house of the Dutch Resident, though its construction would appear to date from the mid-nineteenth century. The 1812 map just referred to does not show the building in its present location, though the site is marked as that of the house of the Resident. The present building is a handsome single-storey residence with a high roof and sweeping lawns in front, guarded by an ancient stone raksasa or giant, from the ruins of a nearby temple.

A little further up Malioboro, on the right, beyond the old Dutch fort and a few houses, is the main market of Yogyakarta, the Pasar Beringharjo. This huge covered market stretches for a long way to the east, with loosely arranged sections selling particular types of goods. The fruits are particularly attractive, and their variety unbelievable; they are neatly piled in pyramids in huge baskets. Batik of course is much in evidence, and prices are often, after bargaining, very low. In the further recesses of the market is a splendid collection of rattan and bamboo products, often finely woven and of excellent quality. The market has nearly everything one could possibly want, and a great deal one would never want at all. The fish and meat section is rather messy but, if one has a strong stomach, worth looking at. The enormous rats which used to be a feature of the market seem to have disappeared.

The tourist information office is located on the same side of Malioboro after the market, near the Kepatihan Danurejan complex, formerly the residence and office of the *patih*, or grand vizier, of the sultan. This is now the administrative headquarters of the government of the Yogyakarta Special Region. Its courtyards and *pendopo*, or open pavilions, are in contrast with functional modern office blocks in the same compound. The Pasar Cina, or Chinese market, is in one of the side streets to the left. Further to the north, on the right is one of the oldest hotels in Yogya, the Garuda, at the level crossing. At this point the road is no longer called Malioboro but Mangkubumi, named after the first Sultan of Yogyakarta. Mangkubumi ends at a road junction marked by a small and rather banal monument, the Tugu, sometimes referred to by Javanese as a needle pointing to heaven.

Along Jalan Malioboro and Jalan Mangkubumi are several examples of Dutch commercial architecture, with stepped gable roofs and windows composed of many small panes of glass on the upper floors, the ground floors mostly having been modernized. The old Dutch residential area was Kota Baru, reached from

Malioboro by turning east at the Garuda Hotel. Before crossing the Code River, on the left of the road, are several rattan furniture shops and then a small flower market. Many of the houses in Kota Baru have seen better days and there are several more recent constructions, but some of the houses retain an air of colonial spaciousness and comfort.

In Malioboro there are two large old-fashioned general stores of Ramai and Samidjaya, on the left and right coming from the south, emporia of a kind much in local favour. Before reaching the Mutiara Hotel on the right from the south is Terang Bulan, one of the best-stocked batik shops in Yogya and run by an old *santri* (orthodox Muslim) family; this has the advantage for some of fixed prices. Almost opposite is the deceptively modest front of Mirota, run by an old *peranakan* (Javanese-Chinese) family; this is essentially a food store cum mini-supermarket, and those who cannot do without Western food and drink will find much that they might want here. There are numerous antique and craft shops the length of the street. By the railway crossing on the right, in the single-storey building, is a high-class batik shop run by Ardiyanto, with many original designs.

Malioboro is a one-way street, running north to south, and a considerable effort has been made in recent years to improve its appearance with the installation of fancy street lights, flower tubs, and so on. It has to carry heavy traffic for which it was not designed and the congestion is often considerable. The siesta is sensibly and fairly rigidly observed by most of the shops and it is useless to attempt to do any shopping between one and five (more often six) in the afternoon.

## East from Benteng Vredeburg

Continuing along the road, Jalan Senopati, from the Benteng Vredeburg, the Bank Indonesia building is located next to the main post office. Further along the same road to the east, past one of the very few Pertamina filling-stations in the city and, behind it, the group of buildings including a church and the Kanisius press, the main Chinese temple in Yogya is found on the right at the next road junction. It has some good wood carving inside.

Carrying on towards the east, when Jalan Senopati becomes Jalan Sultan Agung and crosses the Code River which flows through the centre of Yogya in a north–south direction, the palace of the Paku Alam is found on the left. The Paku Alaman was set up by Raffles in 1813, dividing, as has been mentioned, the principality of Yogyakarta into major and minor rulers in exactly the same way as the Dutch had already divided Solo. The Paku Alam held court in the same way, though on a less lavish scale, as the Sultan, and both in the nineteenth century, having little else to do and little real power, did much to encourage the arts. Eliza Scidmore during her visit had an introduction to the Paku Alam, and left this account of her reception:

> The carriages went through several gateways, past a guard house and sentries, before they drew up in an inner court before an open pringitan, or audience-hall, eighty feet square, whose great, low-spreading roof, resting only on heavy teak columns, was all open to the air. The prince, his crown prince, and his second son, who is the father's aide-de-camp, were waiting to receive us as we alighted....
>
> We went far in over the shining marble floor, away from all glare and reflection of the vast sanded court, to a region of tempered shadow where (the princesses) did the honours most gracefully....
>
> There was a carved screen behind us, closing off an inner space, where broad divans invited to informal ease, and many beautiful objects were disposed. We were taken there by the old prince....The 'Axis of the Universe' is a bibliophile and collector by inheritance....
>
> Conversation never lagged during this morning call....

The kraton of the Paku Alam, the 'Axis of the Universe', is rarely open to the public, but a glimpse inside its gate shows it to be in the form of a large pendopo with related buildings around.

Continuing to the east, the road changes name again to Jalan Kusumanegara. On the right is the state cemetery after which the road is named and where heroes and soldiers of the revolution are buried. The first Indonesian general of modern times, Sudirman, of the revolutionary period lies here.

At some distance from the centre (much too far to walk), after crossing the Gajahwong River, one reaches the Gembira Loka Zoo, on the right. This is a popular place for family outings, and the elephants, in rather cramped quarters, are particularly admired by young visitors. Of interest to Western visitors are two Komodo lizards, thought to be the originals of dragons.

## Monumen Diponegoro

This monument is in Tegalrejo, some four kilometres to the west of the Tugu at the top of Jalan Mangkubumi. It is built on the site of Prince Diponegoro's residence, which was levelled by the Dutch in 1825. The buildings were reconstructed and inaugurated in 1969. There is a fine Pendopo Agung with a good gamelan, and some realistic paintings showing Java's history and the struggle for independence.

Pangeran Diponegoro, as mentioned in the first chapter, was the eldest son of Sultan Hamengku Buwono III and grew up in Tegalrejo where his grandmother, Ratu Ageng, lived. There he meditated and studied Islamic religious texts; he disliked the atmosphere of the court and did not appear there. He underwent a religious experience, and had visions, including a visitation from the Goddess of the Southern Ocean. The Yogyakarta court, after the successful attack on the kraton in 1812, was demoralized and things went from bad to worse in the 1820s, when small uprisings began. The harvest of 1821 was poor, cholera came to Java for the first time. Hamengku Buwono IV died in 1822, rumoured to have been poisoned, and there were well-publicized disputes about who should be the guardians of his three-year-old son and successor, Hamengku Buwono V. Mount Merapi erupted violently the same year.

In 1823 the Governor-General van der Capellen ordered the abolition of all private leases of land in Central Java because of abuses. The princes who had leased their lands lost the source of income and also had to pay back advances and compensate the leaseholders (mostly Chinese and Europeans) for capital improvements. This infuriated the aristocrats. In 1825, when the stakes for a new road to Magelang across Prince Diponegoro's lands near Tegalrejo were being placed, a clash occurred between his retainers and those of his enemy, the patih Danurejo IV. Tension rose, the Dutch sent troops from Yogya to arrest Diponegoro, open fighting broke out, Tegalrejo was captured and burned, and Diponegoro fled, to start the revolt against the Dutch.

For five years he and his followers waged war against government troops and mercenaries. Fifteen of twenty-nine princes joined Diponegoro and nearly half the *bupati* (regents). The heart of the revolt was in Yogya, and the rebellion spread over Central and East Java. Though the court of Surakarta had nothing to do with the matter, it sent troops to help put the revolt down, to little effect. The Sultan also supplied troops to help the Dutch, for Diponegoro was in revolt against the court of Yogya as well, but many troops under Diponegoro attacked Chinese and Europeans, inflicted severe losses and caused a considerable drain on Dutch resources. The period known as the Java War came to an end in 1830 when Diponegoro, his followers deserting, growing weary of the endless warfare, agreed to negotiate with the Dutch in Magelang. The prince was arrested and deported to the Celebes; he died in Makassar in 1855.

With the arrest of Diponegoro the rebellion was over. Eight thousand European and 7,000 Indonesian soldiers died on the government side, and at least 200,000 Javanese died in supporting the rebels. The population of Yogyakarta was thought to have been halved in this period. The revolt has been seen as the last stand of the Javanese elite; Diponegoro stood for his aristocratic rights, and lost. But he stood against the colonial power, and remains a potent symbol of the nationalist movement; copies of his portrait are still on sale in Malioboro and presumably find their way into the most humble homes in Yogyakarta.

# 4
# Arts and Crafts

## Wayang Kulit

THE wayang kulit (shadow theatre) so dominates art forms in Central Java, nowhere more so than Yogyakarta, and so dominates people's ways of thinking, that it becomes almost a way of life rather than an aesthetic distraction. Commentators frequently use the wayang kulit for their analogies to the political scene, to psychological problems, to moral dilemmas, referring to the substance and the shadow of Javanese reality, to characters within the wayang stories, or to the meaning of life as illustrated by the wayang. The wayang kulit engulfs Yogyanese ways of thought; almost everything is seen in terms of the wayang.

The origins of the wayang kulit are obscure. It may have come to Java, as the stories most commonly presented in the wayang, the *Mahabarata* and the *Ramayana*, certainly did, from India along with the religious and cultural influences to be found in Java in the seventh, eighth and ninth centuries. It seems likely to have been well-established in Java at the time of Majapahit, though again this is not entirely certain (the term wayang can cover any theatrical form). The use of leather for the figures is traditionally ascribed to a king of Demak in the early sixteenth century and the extreme stylization of the figures is said to be due to the Islamic taboo on the representation of human forms; neither can be proven (this said, the shadow play figures in Bali, where many Javanese pre-Islamic traditions have been preserved, are more realistic than those found in Java).

An earlier form of the wayang kulit was the *wayang beber*, when the stories of the wayang were recited by a narrator who unrolled painted paper scrolls illustrating his tale. This form certainly existed in the fifteenth century but is now almost completely extinct. Some scrolls survive and the figures on them are very similar to the wayang kulit figures; however, the popularity of wayang kulit may have led to wayang beber figures being painted like the wayang kulit.

A variant of the wayang kulit is the *wayang klitik*, where the figures are made of flat carved painted wood with leather arms. This form was popular in East Java but has largely disappeared now.

The wayang kulit figures are made of different kinds of leather, the most esteemed being water-buffalo skin. The leather has to be treated so that grease is removed, and to make it sufficiently thin it is beaten and dried in the sun many times. The figure desired is outlined and then cut with a fine knife. Chisels and hammers are used to obtain the cut-out effects. The moving arms are attached before the painting process, and traditionally only natural colours were used. The figures were then covered with a thin glazing to prevent chipping. Lastly, the figure is fastened to the *cempurit*, the sticks made of horn, for holding the figure upright and for moving the arms. The horn is usually taken from a water-buffalo, the best being that of an albino animal.

A full wayang kulit set is likely to consist of some 400 figures. Before a performance they are ranged to the left and right of the screen in decreasing size, the smaller ones towards the centre. This means in practice that the giants and ogres, hairy, coarse and gross, are at the extremes and the more refined and elegant characters, slender and withdrawn, in the middle. They are held in place by their horn sticks stuck into two banana tree stems placed at the bottom of the stretched cotton screen.

The *dalang*, or puppeteer, sits cross-legged in front of the screen, with the figures he is likely to use for the episode chosen for the performance close to hand. To his left is the wooden box in which the figures are kept, and he hits some metal plates hung from the box with the *cempala*, a wooden mallet, which he activates with his foot. The rhythms he produces have a double effect; they intensify the drama at particular points and give cues to the gamelan orchestra which always accompanies a wayang kulit performance. Above his head is a lamp, originally an elaborate oil lamp made of bronze, but now more likely to be a strong electric light bulb with a zinc cover on one side and projecting the light towards the screen. The performance may be watched from behind the performers, as is now more common, thus seeing the figures in full light with some shadow beneath, or from behind the screen, seeing the figures only in shadow.

The gamelan may consist of thirty or so musicians. The most important instruments are the *kendang*, or drum, which sets the tempo and gives signals for beginning and ending pieces, and the *gender*, a gentle metallophone, which sets the moods for the different scenes and the musical modes and pitches, playing therefore throughout the performance. The *rebab*, a stringed instrument, and the singers, *sinden*, are also important since they introduce the pieces the orchestra will take up and develop.

But the dalang, who will perform without interruption from about eight in the evening until dawn breaks the following day about five, is the undisputed master of the show. He selects the story and the characters, and develops both; he sings, speaks, manipulates the puppets and directs the gamelan orchestra, telling the musicians when and what to play. Each story is known in outline only, and so he must be familiar with a very large repertoire. He improvises the development of the story as he goes along, adding humour, pathos and drama where he pleases. He manipulates the gods and their royal progeny. Small wonder that the dalang is a revered personage in Javanese society and his years of training lead to comparatively high rewards.

The Ramayana stories develop episodes relating to the hero Rama and his brother Leksamana, the abduction of Rama's wife Sinta (Sita) by Dasamuka (Rawana), the assistance of the monkey leader Anoman (Hanuman) in securing the release of Sinta from Ngalengka (Lanka), and the final battle between the warring factions. There are numerous variants and additions to this basic story

The Mahabarata stories are performed much more frequently. The complications of the family rivalry between the Pendawa and the Kurawa would require volumes to describe: basically, the Pendawa may be said to have right on their side and the Kurawa to be in the wrong, but to say that is a gross over-simplification and the truth is far more complex. Black and white are never so precise in Java or anywhere else and there are good and bad on both sides. Arjuna is the most refined of all the characters in this cycle, Bima is powerful in battle and can fly, his son Gatutkaca is less gruff than his father but as powerful. Sumbadra, one of Arjuna's wives, is considered an ideal of Javanese femininity. All the characters are instantly recognized by the audience, aware of the different physical appearance of each puppet and the personality of the character it represents.

Mention needs to be made of Semar, the clown, with his three sons Petruk, Gareng and Bagong; these have enormous importance and represent far more than the servant world and foils to their royal and noble masters. They embody truth as well, they give advice and comfort, and though grotesque in appearance, are nonetheless godlike. Lastly, there is the figure known as kayon, of which mention has been made earlier, and which marks the beginning and end of the performance and changes of scene; this does not represent a human form, but a stylized view, full of symbolism in its images, and in the shape of a mountain.

There is a school for dalang in Yogya and wayang kulit performances are frequent. Western visitors often get bored after an hour or so, and it is better to go about midnight when more interesting battles between the giant armies take place, or at three in the morning when the shift to the dénouement occurs. It is preferable to go with someone who is proficient in Javanese and who can explain the action and the words of the dalang and the singers. It is difficult for visitors to understand the importance of the wayang in Javanese lives, but to watch the interest and anticipation of the spectators at a performance will give some idea, as will the exquisite intricacy of the music and the subtle brilliance of the dalang's art.

## Wayang Golek

*Wayang golek* is another puppet theatre, performed with three-dimensional realistic wooden stick puppets, elaborately costumed and usually with an appropriate head-dress. There is a central rod which passes through the body and turns the head, so that the shoulders can be moved independently of the head, to add realism. The arms are moved by separate sticks. Each figure is about twenty to thirty centimetres high. Wayang golek is shown on a miniature stage, complete with curtains; the puppeteer, who for the most part cannot be seen by the audience, is seated below and slightly behind the stage, and manipulates the puppets by holding them in the air at the level of the stage opening.

Most of the stories are derived from the Mahabarata or from a Javanese compilation, *Serat Menak*, about an Arabian prince, Amir Hamzah, an uncle of Mohamed. In spite of this, the costumes are often of Javanese court dress of the eighteenth century, with the men sometimes wearing eighteenth- or nineteenth-century frogged jackets; some characters are in Arab dress, though rarely in Yogya. All have a long batik skirt gathered at the waist to hide the central rod and the puppeteer's hands holding it.

The Menak stories of the wayang golek are thought to have been introduced to Java fairly late and probably by the people who brought Islam to the northern coastal ports. However, the actual form of the puppet seems to be a Javanese adaptation of Chinese hand puppets or string marionettes; this is not so unlikely when one remembers the existence of long-established Chinese communities on the north coast. The use of rods in the wayang kulit would have been transferred to the string marionettes.

Wayang golek is far more naturalistic than wayang kulit; the figures can fume in rage and pant with exhaustion as the puppeteer repeatedly raises and lowers the trunk of the body on the central rod. The facial representation is also very naturalistic, with few of the conventions of wayang kulit (however, slit, half-closed eyes indicate nobility and grace, soya bean-shaped eyes are for intermediary personages, and round bulging eyes are for demonic, crude or violent characters as in the wayang kulit). The court head-dresses and tall black fez banded with gold worn by patih are impressive and elaborate.

There is much less ritual and importance attached to wayang golek and the performances last nowhere near so long. They are reasonably popular in Yogya and a couple of villages nearby specialize in wayang golek. In West Java, however, the Sundanese have a more elaborate wayang golek, with stories from the Ramayana and the Mahabarata, and there the wayang golek takes precedence over wayang kulit.

## Wayang Topeng

*Wayang topeng* is an ancient form of masked dancing, now fairly rarely performed in Yogya, though still common in Bali. It appears to have been popular at the Majapahit court and is mentioned in a fourteenth century text, the *Nagarakertagama*. It may have fallen out of favour under Islamic influence but the dramas were revived in the eighteenth century courts of Solo and Yogya and flourished until recent times.

The shape of the mask is like an inverted pear, almost triangular. The same conventions for the shape of the eyes in wayang kulit and wayang golek apply. Elevated (alus) or vulgar (kasar) personages have different kinds of masks, which are painted and sometimes gilded. Whatever the size of the eyes, the actual slits for the dancers to see through are very narrow, and they have to hold back their heads and move them gently, even if the rest of the body is still, in order to have an adequate field of vision. The mask usually has its lips slightly parted, so the characters appear to have an elusive smile. Old masks have a strip of leather on the inside near the mouth for the dancers to grip the mask with their teeth, but newer masks do not have this, the modern practice being to strap the mask to the head. Some masks, usually for the coarser characters, are in fact only half-masks, stopping below the upper lip and lacking the pointed chin common to the more refined characters.

There is no scenery and no theatre; the dancers, usually men, perform on an open floor of a pendopo to the accompaniment of a gamelan orchestra, as for the wayang kulit. The costumes are, however, elaborate and closely correspond to those adorning the characters in the shadow play. Wing-shaped ear ornaments, tassels, crescent-shaped plates on the chest, armlets, and bracelets are worn; the men have knee-length tight trousers covered by a huge batik sarong. They wear a long sash, or *sampur*, around the waist with two free ends falling to the floor; a great deal of the technique in Javanese dancing lies in the way these scarf ends are used. They are flicked to indicate disdain or impatience, held out to show modesty or hesitation, and so on.

The stories are, as in the wayang kulit, usually taken from

episodes in the Ramayana and the Mahabarata, or more often from the Majapahit *Panji* cycle of stories. The dancers now speak or sing their own parts, in classical Javanese; formerly a dalang acted as narrator.

## Wayang Wong

*Wayang wong* is very close to wayang topeng but masks are only worn by giants and animals: the natural faces of the human characters are perfectly visible. The costumes, the setting, and the musical accompaniment are virtually the same as in topeng. The stories come from the Mahabarata, the Ramayana and the Menak cycle. The wayang wong (wong in Javanese means 'people') reached a peak at the four courts of Central Java, and the two courts at Yogyakarta strove to outdo each other and those at Solo in the perfection of their dance-dramas.

There is one other important difference, however, between the wayang wong and the wayang topeng; the wayang wong may have a theatrical backdrop and it is occasionally performed in a standard theatre with a proscenium arch.

In dancing style, the Yogya wayang wong dancers bear a close resemblance to the wayang kulit figures they are dressed like. They present their faces in profile to the audience and their gestures are angular and precise rather than fluid and human. The dancers' roles are determined by their stature; if the men are tall and well-built, they will have the roles of forceful characters like Bima, demons or monsters, but if they are slender, of average height, with almond-shaped eyes and a pointed nose, they are cast in the role of refined characters. The men are costumed as in wayang topeng; the women wear long batik sarongs often trailing on the floor between their legs and kept in place by a swift movement of the feet. Emotions are expressed with extreme restraint, being most often exteriorized in the use of the dancing scarf, the sampur. Faces tend to be expressionless, even mask-like. The clowns exist in both wayang topeng and wayang wong and of course derive from wayang kulit.

As in wayang kulit, wayang wong and wayang topeng, stories are known only in outline. The dancers also have to act, speak and sing, using the language of classical Javanese, delivered in extremely stylized cadences, much more so than in topeng. It is the duty of the dancers to improvise their verbal roles, which they usually do on stage.

There is an increasing tendency for women to play the alus male parts in wayang wong, leaving only the 'strong' roles for the male dancers.

## Classical and Sacred Dance Cycles

There are dances performed outside the conventions of wayang wong and wayang topeng. These are particularly the sacred and court dances like the *bedoyo* and the *serimpi*, with stories drawn from the Mahabarata, the Ramayana, the Menak tales, Javanese folklore and Javanese history. They are performed by exquisitely costumed young girls in fairly rigid body positions, moving only the hands and fingers, gently rotating the head, looking modestly at the floor and gliding on tiptoe across the floor in subtle movements which seem barely perceptible.

Practice is systematic and was originally maintained in the courts. At a practice, the dancing master acts like a dalang of the wayang kulit, knocking on the *keprak*, or slit wooden box, with a mallet to give signals to the dancers for their next movement.

Most Westerners find classical Javanese dancing extremely slow and its subtleties too refined. Balinese dancing, being more popular in origin and altogether more lively, is much more accessible. However, classical Javanese dancing is a stately product of refined courts, and it takes time to appreciate its particular qualities.

## Ketoprak

*Ketoprak* is a popular form of theatre with stories drawn from folklore, history and popular cycles, including the *Damar Wulan* tales about a youth employed as a grass cutter at the Majapahit court and who ends up in marrying the maiden queen and gaining the kingdom. This form of folk melodrama is performed in Javanese and the actors improvise freely, especially with the jokes, which are often verbal puns. Ketoprak seems to have originated in the

Yogyakarta-Solo area in the First World War period and to have become a craze by the 1920s: because of language difficulties, it is not greatly appreciated by visitors.

### Sendratari

*Sendratari* is a modern dance form, basically a dance theatre without words and much inspired by modern Western ballet. The length of the pieces performed is more in line with contemporary fashion and unnecessary action is eliminated. Choreography is an important consideration in this very fluid and quite popular form.

The most successful and certainly one of the most spectacular of the new sendratari productions is probably the Ramayana ballet, performed at Prambanan on four consecutive nights of the full moon in the dry season, with hundreds of performers.

### Music

The composition of the gamelan orchestra has already been described earlier in this chapter when dealing with the wayang kulit. All the instruments are percussive with the exception of the stringed rebab, the zither-like *celempung*, the flute (*suling*) and of course the sinden, the singers. In spite of this, the sound of the gamelan is incredibly delicate and gentle, it having been compared to rippling water and moonlight. Sir Francis Drake, when he heard the 'country-musik' of 'Raja Donan, King of Java' in 1580 when the *Golden Hind* called at the south coast of Java, noted that 'though it were of a very strange kind, yet the sound was pleasant and delightfull'. Raffles was so impressed by the sounds of the gamelan that he brought no less than two complete orchestras back to England with him.

There are most often eighteen basic instruments in the gamelan, of which two are drums and four single gongs of different sizes. The gongs and the *bonang* are probably the easiest to play. The latter consists of a rack of bronze bell-like objects, hit with two padded sticks. The wooden and metal xylophones, on extremely elaborate stands, are struck in the same way. The gender, which leads the music in the wayang, a xylophone strung over resonators, is the most delicate in sound of all the gamelan instruments and the most difficult to play; as two notes (rarely in unison) are being struck with padded hammers, the two preceding notes have to be dampened with the thumbs, so that one is playing four different notes at the same time. The different gongs mark the end rather than the beginning of sections.

All gamelans have two different, non-interchangeable tunings; pelog, which has seven unequal distances in the octave, and slendro, five almost equal distances in an octave. This means that to play any piece within the Javanese repertory one needs two sets of instruments.

The music cannot be written in Western notation but a simple numbering system is sometimes used when music is being learnt, each note on an instrument being accorded a number. The timing would be indicated by the drum. But most music is learned by heart, without any recourse to writing systems, and musicians from the first few notes of the lead played by the gender can immediately pick up what they are supposed to play and carry on. Most people can play on any of the gamelan instruments as required, and can change or move from one to another during a performance. Players of most instruments are usually men, with the exception of the gender.

The basic theme is usually quite simple and its development one of continuing expansion and intricacy. Playing and listening is considered by many Javanese a spiritual discipline. It is also a form of group activity for, with the exception of the gender, it is virtually impossible to play any one instrument in the gamelan as a solo instrument.

Raffles, in his *History of Java*, noted:

> . . . it is the harmony and pleasing sound of all the instruments united which gives the music of Java its peculiar character among Asiatics. The sounds produced on several of the instruments are peculiarly rich, and when heard at a distance have been frequently compared to those produced on harmonic glasses. The airs, however simple and monotonous they may appear of themselves, when played on the *gambang kayu*, or accompanied by the other instruments, never tire on the ear, and it is not unusual for the *gamelan* to play for many days and nights in succession.

The gamelan is Java's unique contribution to the gong-chime cultures of East and South-East Asia. The Balinese gamelan is derived from the Javanese model, but is used with more overt dramatic and sometimes deafening effect. The Javanese gamelan is subtle to an overwhelming degree. It is also extremely ancient; some of the instruments can be seen in more or less their present-day form in the carvings at Borobudur. However, it has not fossilized; the stringed rebab is an instrument of Arab origin and was probably only added in the sixteenth century, and the addition of male and female singers is a fairly recent development of this century.

Gamelan music accompanies every theatrical representation to be found in Yogyakarta (the East Javanese popular theatre form, *ludruk*, which has no gamelan, is hardly ever seen in Central Java) and all classical as well as most modern dance forms. Gamelan orchestras sometimes play by themselves, without any accompanying visual entertainment. Even for the non-musical, the visual aspect of the musicians in traditional court dress and the exotic nature of the instruments is fascinating, and the delicacy of the music itself cannot but make an impression.

## Batik

As with the gamelan, on which whole books have been written, batik is a complex art which is at the heart of Yogyanese culture and an enormous subject in itself.

The art of batik is extremely ancient, though when precisely it was introduced, presumably from India, is uncertain. The technique is complex and time-consuming. A design is traced on white cotton, and the areas to remain white are covered with wax. The wax is applied with a *canting*, a small tracer with a spout and a container for the wax. The wax is kept warm over a small brazier heated by charcoal. Only small amounts can be applied on the cloth at any one time because the wax soon gets cold and stops flowing through the tiny spout of the tracer. The canting is a refinement possibly introduced in the sixteenth century, the wax being originally applied with bamboo tracers.

When all the parts to remain white have been covered with wax, then the cloth is dipped in dye and dried. The whole cloth, except for those parts covered in wax, now has the colour of the first dye. Wax is reapplied to those parts which are to keep this colour, and then the material is dyed again in a different colour. More wax is applied to those parts to keep this second colour, and so on, until the process is complete. The colours used at first are pale—yellow, light blue—and those at the end of the process dark—deep brown or black. The wax is removed by boiling and scraping.

Concentration and patience are required to apply the wax, for the slightest error ruins a whole piece. Traditionally the art of tracing the designs and applying the wax was reserved for court ladies. Certain designs were the preserve of the nobility, like the *parang rusak*, or broken kris design, but this is no longer the case. Most pieces were sarong lengths, two and a half metres long by one metre wide.

Traditionally colours were natural, indigo providing blue and the bark of certain trees the deep brown, and limited in range. Now chemical dyes are used and the range is very wide.

A single piece of fine design may take many months to complete and so is comparatively costly. The process has been speeded up in some cases by the use of metal printing blocks called *cap* which can repeat a small design of roughly eighteen by twenty centimetres. This reduces both the price and the effect of the piece produced. Completely industrialized processes are being introduced using batik designs with silk screen-printing methods, but such lengths cannot be considered as true batik and the result is obviously inferior.

Yogyakarta is considered the home of batik, along with Solo, and there are many outlets for the material in the city. The quarter of Tirtodipuran, to the south of the kraton and next to Prawirotaman where there are numerous guest-houses, is the traditional centre of domestic batik production. A number of small-scale factories producing cap lengths are also found here.

A variant of cloth lengths for clothes has been the use of batik as wall decoration. Some of the traditional lengths make acceptable pictures, even if not so intended, but batik pictures have taken off as a peculiarly Yogya art form, and they are on sale everywhere.

Unfortunately, there is a great deal of copying of the more successful artists going on, with too little attempt at originality. There are, however, some very good artists indeed using this medium.

## Oil Painting

Painting in oils was a Dutch innovation in Indonesia, though painting on cloth or paper, probably similar to that now still practised in Bali, existed in the later Majapahit kingdom. With the notable exception of Raden Saleh, who lived a long time in Europe and was influenced by Delacroix, painters in the nineteenth century were mostly Dutch, producing romantic scenes as souvenirs for colonial planters and the like. The Taman Siswa schooling system, on the Montessori model, which originated in Yogyakarta in 1922, laid stress on the arts and a number of talented Indonesian artists emerged in the late 1930s and early 1940s. The heyday for painters was the period of independence and revolution and Yogyakarta, being the capital, was the centre of artistic activity. The Yogya school laid stress on social realism and President Sukarno, a great collector of oil paintings, encouraged the 'Indonesianism'—as opposed to Westernization—of local painting. The existence of the state Fine Arts School in Yogyakarta was an added stimulus. But the increasing politicization of oil painters and their subjects in the late 1950s and early 1960s led to a decline in quality; furthermore, the centre of attention began to shift to Jakarta, particularly in the 1970s with the creation of the Taman Ismail Marzuki. However, a number of talented painters remain in Yogyakarta, of which Affandi and Saptohudoyo are probably the best-known. There are many others, though they are less obvious than their batik painting counterparts, and they are in many ways much more difficult to locate. The possession of an oil painting remains the hallmark of bourgeois respectability. Some perfectly hideous ones, showing erupting volcanoes and dark river valleys, are hung up on the railings on Jalan Solo by the bridge over the Code River (from which, incidentally, those indebted by gambling or ill-starred in love throw themselves in despair). They presumably find buyers.

## Leather Work

Wayang kulit figures are the most traditional art form in leather in Yogya and are made in several ateliers in the city. The process of making these has been described in the section on wayang kulit. The same process is also used for producing coloured and gilded leather lampshades which may be an adaptation of a traditional craft but cannot be considered particularly satisfying aesthetically.

Yogya produces a great deal of more practical leatherware, in the form of suitcases, handbags, sandals, and belts. There are a couple of small factories which make to order leather overcoats and jackets, as well as shoes, for those returning to cold climates.

## Metal Work

The most esteemed metal work being done in the Yogyakarta region is the silverware at Kota Gede, with its elaborate designs and embossed patterns. Metal work is also produced in relation to gamelan instruments. On the whole, with the exception of the silver, production is utilitarian rather than decorative.

## Pottery

In the village of Kasongan, to the south of Yogya on the Bantul road (it is the second road on the right past the Madukismo sugar mill) are several potters making very attractive and often amusing ornamental figures and pots. They excel in producing fantastic animals, horses, tigers, lions, elephants, birds, tortoises and so on. The dark red clay is obtained locally and the craftsmen are entirely untrained, usually being young men helping the family. The prices asked are very reasonable.

Unfortunately, the work is only lightly fired, and so is rather brittle. It is possible to request that items be double fired, but the colour then becomes a dull grey or black. Sometimes the pieces are painted, but the colours used are gaudy and the natural earth colour is more pleasing.

The short journey to Kasongan, which is very close to the city, is worth it not just for the pottery. The village itself is quite typical and unspoilt, a pleasant oasis of peace in the rice fields.

# 5
# The People

The people of Java, according to Geertz' classic distinction, can be divided into three classes: the *priyayi*, or aristocrats, much given to mysticism, the *santri*, or devout Islamicists, often though by no means always traders, and the *abangan*, comprising the vast majority (the *wong kecil*, 'the little people', in Javanese), with their folk beliefs and humble occupations. But Geertz was discussing the Javanese in terms of religion, and it would be unfair to ascribe to him a social mould which appears rigid. The entire social range of Yogyanese society cannot be neatly labelled and divided into three; the truth is far more complex. The gamut from the sultan to the humblest rice farmer or street beggar has enormous variety and, like many traditional societies, the Javanese have very fine distinctions between different levels of society.

These different levels are categorized by, among other things, language. Javanese has three major levels of language, each with its own vocabulary, and particularly a wealth of personal pronouns, nouns and verbs; Geertz distinguishes two variant levels, making five in all. The person speaking uses the level appropriate for the person being addressed; if that person is of the very highest rank, then the highest form, *kromo inggil* is used; the next highest is *kromo biasa*, or just *kromo*. The level below this, middle though still elevated, is *kromo madya*, or just *madya*, and is used between persons of equal rank and status (something very rare in Javanese society). Low Javanese is known as *ngoko*; Geertz distinguishes two variants of *ngoko*, both only used by the priyayi class when addressing people of lower rank, and called *ngoko madya* and *ngoko sae*, or *ngoko alus*.

Javanese society strives to present to the world an outwardly calm, dignified and refined (alus) aspect. To do otherwise is to be kasar, or uncouth; nothing is more looked down upon than such conduct, which disturbs the order of society and *rukun* (harmony). These priyayi values are underlined by two further concepts, of *lair*, or external human behaviour, and *batin*, internal feeling or inner life; the former, always deferential to others, conceals the latter.

The aristocracy in the nineteenth century was increasingly absorbed into the administration since it had nothing else to do. The enormous increase in the bureaucracy in the early years of the republic, with, as is common in such cases, a corresponding diminution in actual productivity, led to a considerable thinning of priyayi values, yet they nevertheless prevail, nowhere stronger than in conservative Yogyakarta. The aim of most urban families is to install their offspring in the bureaucracy; the work is undemanding, the hours leisurely (government offices are open usually from eight a.m. to two p.m. only, sometimes less, and on Fridays close at eleven a.m.), the salary, if extremely low, assured, and at the age of fifty-five a pension will be forthcoming which, no matter how insufficient, will be better than nothing. The restricted hours of government offices allow the more enterprising to take on other jobs, and there are often perquisites of office which allow a small and necessary increase in one's income.

The ramifications of Indonesian bureaucracy are considerable and the number of government offices and agencies huge. Yogyakarta is a *daerah istimewa*, or special district, with the status of a province, in which the sultan acts as governor. This means that every branch of the central government has to be represented at the local level.

Yogyakarta is a major educational centre in the country, and all teachers, lecturers and support staff, with the exception of those in private schools and colleges, are state employees as well. This emphasizes the role of the government as employer in the special district.

At the apex of the educational pyramid is Universitas Gadjah Mada, founded in 1946 and originally housed, as has been mentioned, in the front part of the kraton. It currently numbers some 20,000 students and nearly 1,000 teaching staff, nearly all of whom moonlight in other institutions in order to gain a living wage. It is now located in buildings not notably original in their architecture to the north of the city in an area known as Bulak Sumur, though the teaching hospital remains in the south of the city. Gadjah Mada, as the oldest and second largest university in the country, is a nationally important institution, recognized as such. Students come from all over the country, though as might be expected, given that the overwhelming majority of Indonesians are from Java, Javanese dominate.

The advantages of Yogya as a university city are its distance from the inevitable politicization of the capital, its relative cheapness, and its calm. Students, therefore, come from the large islands of Sumatra, Kalimantan, Sulawesi, and from nearly all the smaller islands as well as the province of Irian Jaya.

Close by Gadjah Mada University (UGM or Gama as it is often known), in the same part of the city, are the state degree-giving teachers college, IKIP Yogya, and the private (Catholic) degree-giving college, Sanata Darma. Not far from these is the private Universitas Islam, on the road to the airport. All three have students from all over the country and the Islamic University has students from other countries as well (including Thai Muslim students).

Also by Bulak Sumur, off Jalan Gejayan like the two IKIP, is ASTI (Akademi Senitari Indonesia), the state dancing school, founded in 1964. Three types of dance are taught here: classical Javanese, of course, both Yogya and Solo variants, Balinese and Sundanese. The training is rigorous and the public performances much esteemed.

ASTI is not the only state academy of the arts, though it is perhaps the most famous. ASRI, to the west of the post office, just beyond the bridge over the Winongo River and to the right, teaches fine arts, notably painting and sculpture. Batik work is also done here, but batik is so much a daily tradition in Yogya that it hardly has to be taught in the art schools.

The state music academy, ASMI, is to the south of the kraton and Plengkung Gading, slightly to the west off the road leading to Krapyak. This teaches classical Javanese and also Western music. It is perhaps not always realized outside of Indonesia that Western classical music has a number of devotees in the country, as can be seen from the large number of cassettes of such music available in the shops.

There are numerous private academies of this and that, several private universities of mostly lesser standing and any number of language schools. Primary and secondary schools, both state and private, are everywhere. In 1973 there were in Yogya 195 junior and senior high schools and 47 universities and academies, and the number is certainly not less now. Yogya is a student city par excellence, and the favourite mode of transport remains the bicycle, though this is increasingly being taken over by the motor bike, on the back of which girl-friends balance side-saddle in a precarious manner with complete nonchalance.

Although the major educational institutions are concentrated in the north of the city and such *asrama*, or students hostels, as exist officially are mostly to be found there, students in fact are lodged, as are their teachers, all over the city. Many houses rent out a room or two to students, quite often for a modest sum, in return for some tuition for the children of the house.

With the presence of so many students, literateurs and artists, Yogyakarta is perceived as an intellectual centre of importance in Indonesia. The period of the revolution saw a great deal of literary activity. The most famous writer living in Yogya is probably Rendra, usually considered Indonesia's greatest living poet, and a dramatist and theatrical producer of considerable talent. The grand old man of Indonesian painting, Affandi, lives in a striking house in Jalan Solo on the way to the airport and a number of his paintings are displayed there. Two of the most successful batik artists in Indonesia, Bambang Oetoro and Amri Yahya (who in fact comes from Sumatra) live in Yogya, as does the sculptor-artist Saptohudoyo who now has a gallery near the airport in an extraordinary building like a concrete central American temple.

There are numerous small, often ephemeral, galleries and shops serving as outlets to the artists living in Yogyakarta, but more humdrum trade is carried on widely. Traders exist of all kinds, from the successful Chinese enterprises, usually found in the main street, Malioboro, but by no means confined to that part, to the humblest *warung*, or stalls, on the street corner, lit at night by a small kerosene lamp or a flickering coconut oil flame. There are relatively few factories in Yogya; there is a milk canning establish-

7 Krapyak: Adukidang

8 Borobudur: view from the top circular terrace

1 Kraton: front facing Alun-Alun, with becak

2 Kraton: Plateran Kedaton, main courtyard

3 Kraton: Gedung Purworetno, Sultan's office

4 Kraton: Plengkung Gading, southern gate

5 Taman Sari: Gedung Cemeti and the restored lanthorns

6 Taman Sari: Sumur Gemuling, stairways

9 Candi Mendut: external view

10 Candi Prambanan: view from the north-east

11 Candi Kalasan: detail

12 Candi Sambisari

13 Candi Sari: view from the east

14 Candi Sewu: guardian raksasa

15 Candi Plaosan: view towards Merapi

16 Dieng: Candi Bima, tiered roofline

ment, a cigarette factory, some copper and ironware manufacturing, and a number of cloth weaving establishments. Large-scale industry is unknown. The Chinese are not numerous in Yogya and maintain a very low profile outside the commercial streets of Malioboro and Jalan Solo. There are a number of traders from Minangkabau (they seem to have cornered the antique market, for example), but the majority are undoubtedly Javanese.

At the lower end of society are the night hawkers, like the men selling *tikar*, or rush mats, by the railway station, or the sellers of brooms and brushes who roam the residential quarters of the city. There are the cigarette butt collectors, who go around picking up with long tongs the stubs thrown down by those with money to spare; the butts are unwrapped, the tobacco taken out and rerolled in fresh paper, and sold individually in wayside stalls. There are the itinerant musicians who move around, two or three together or sometimes singly, playing gender, metal xylophones, or zithers, sometimes singing too, and were sent away with a five rupiah coin (with inflation, this is now more likely to be Rp25). Other street entertainers include the *jatilan*, the 'wild horse' performers, when one whips the other to activity, both entranced. And there are the beggars, a considerable number of them, of all ages and both sexes. The 'rent-a-baby' beggars are invariably female, sometimes old, sometimes young, dressed in suitable rags; some are genuinely destitute, others are professional beggars who do quite well at their trade. Any number of infirmities are paraded and charity is sought to counteract them.

The *becak*, or pedicab, drivers form a class by themselves. In 1975, according to the city police, there were 5,971 becak, which probably means that in fact there were more, since a number are bound not to have been duly registered. Since most becak are rented out, with less than ten per cent being owner-driven, the actual number of becak drivers, who work in shifts on a single vehicle, must be at least 15,000. Most becak drivers are young, in their twenties, with, according to one survey, more than half of them never having been to or completed primary education and more than half having previously been farmers. Most became becak drivers because no other job was available and they would prefer to do any other job than be a becak driver, for the work is hard and the rewards few. The majority want their children to be government employees. Occasionally one meets becak drivers who are students earning money to see them through school or even university, but this is very rare.

The becak is, or used to be, the most frequently used mode of transport within the city for both goods and people. As however the city has spread and the number of individual owners of bicycles and motor bikes has increased, as an internal bus service of a sort was introduced in the city, the actual use of becak may not have increased in proportion to the population increase and may even have declined. Becak are slow and sometimes inconvenient, especially in wet weather (when, of course, bicycles and motor bikes are no less inconvenient). Accidents are remarkably few, though close shaves frequent. Many Westerners regard taking a becak as degrading to the individual; the becak driver, however, is happy to get customers who will cover the cost of his next meal.

As well as becak, there are two other distinctive forms of transport in Yogya. These are the *gerobak,* or oxcart, slow, lumbering vehicles on rubber tyres and attractively decorated, which are used to transport heavy loads, particularly of construction materials, usually drawn by two oxen, and the *andong,* or horse-drawn garry, still a popular means of transport over distances too great to be covered by a becak and with a larger number of persons than can be accommodated in the pedicab. Andong are particularly used in the areas outside the city centre and for transport between villages.

A number of occupations are plied in Yogya which are typically Javanese. The *bong supit* are the traditional circumcisers who are paid to perform their operations by grateful families of boys aged usually about twelve years old; traditional circumcisers are reputed to use magic and hurt less than doctors in hospitals. They also cut less skin. The *dukun,* or sorcerers, are much used by most Yogyanese for advice and for healing. They sometimes work as a husband and wife couple, with the wife assisting the husband to go into a trance and speak with the voice of spirits who reveal the unknown. There are dukun of all kinds, but those most in demand are the curing dukun who remove the 'air' or 'heat' from inside the body and with them the ills they cause. In most families a healing dukun is consulted as often as, if not more frequently than, a doctor in case of illness. Sorcery is widely believed in and is in the hands of specialist dukun. Particularly strong emanations from the spirits are obtained in graveyards on Thursday evenings and to consult a dukun there at such a time means one must have a very strong constitution or be in great danger from the other world.

The spirit world for most Javanese is a real one; Geertz divided the spirits into five main categories, *memedi*, or frightening spirits, *lelembut*, or possessing spirits, *tuyul*, or familiar spirits, *demit*, or place spirits, and *danyang*, or guardian spirits. Some are friendly, most are not. The *sundel bolong*, a prostitute with a hole in her back hidden by her long hair, is one of the best-known frightening spirits, though the most common are *gendruwo* who play tricks on people. Tuyul can be friendly to their owners, for they steal on their behalf, and make the rich even richer, but of course they are a curse to those they steal from; tuyuls are said to be common in the batik trade in Yogya. The most powerful and widely-acknowledged spirit in the Yogya region is undoubtedly Ratu Loro Kidul, the Goddess of the Southern Ocean, with whom all the rulers of Mataram have been in special communion.

The spirits can be placated by the dukun directly or on his advice (often involving going to crossroads at midnight) or with his aid (drinking potions with a variety of ingredients). Or they can be placated by a traditional *selamatan*, or ritual meal, which is the classic Javanese way of marking an occasion, with particular foods arranged in special ways according to the occasion being celebrated. The marking ceremonies of life, that is birth, circumcision, marriage and death, are all celebrated with selamatan, the first and last at fixed intervals in terms of numbers of days (10, 50, 100, 1,000, and so on) after the event.

The religious views of the inhabitants of Yogya represent, for the most part, a complex intermingling of traditional spirit beliefs, residual Hindu-Buddhism which prevailed from the seventh to the sixteenth centuries, and a veneer of Islam. In addition, there are many converts to Christianity and several Catholic and Protestant churches. In the elections of 1955, when the vote was along religious as much as party lines, Yogya had the lowest percentage in the whole of Java voting for the Islamic parties, only twenty-five per cent. If anything, the hold of Islam has possibly lessened since then, for after 1965 everyone was required to state a religious belief (on the theory that with one a person could not be a communist) and many opted for anything except Islam. This said, the vast majority of Yogyanese are nominal Muslims; they may never go to the mosque, they may eat pork or drink beer, they may not fast during Ramadan, but they observe the rites of circumcision and burial.

A particularly Javanese, more especially Yogyanese, religious variant is belief in mysticism and the practice of *kebatinan*. The metaphysical vagaries of the various mystical sects are numerous, but their power and influence cannot be overlooked. A few years back there was a move on the national level to have mysticism recognized as one of the official religions of the state. The Yogyanese aristocrats (priyayi) have developed mysticism, with attendant meditation, breath control exercises, concentration of one's powers to a single end into a habit, if not a creed, clearly inspired by elements of Buddhist practices. Along with this development of the mind goes the relative disinterest in the body and the worldly life; fasting as a form of self-denial is practised, and the Javanese remain comparatively unmaterialistic. By and large and with obvious exceptions, they do not adorn their houses, their dress is subdued and simple, their cuisine modest. This may in many cases be no more than making a virtue out of necessity, but it also reflects a Yogyanese frame of mind; inner harmony is more important than externals.

As in most of Indonesia, politics in Yogya since the events of 1965 have played no significant role. The Sultan has now retired from activities of state, the dominance of the bureaucracy as employer is reflected in the strength of the pro-government army-administration 'functional grouping' Golkar. There were a few incidents between students and the army in the late 1970s but these were quickly suppressed. Yogyakarta has been relatively quiet in recent years, noting mostly an increase in students and tourists.

This said, Yogya remains one of the poorest and most over-populated regions of Indonesia and the need for jobs, particularly among the young, is chronic. For this reason many of Yogya's youth go to seek their fortune elsewhere, particularly in the capital. Yogya is the place they will dream of retiring to, where life is quiet and peaceful and living cheap.

# 6
# The Ancient Monuments

## The Borobudur Group

*Borobudur, Mendut, Pawon*

BOROBUDUR, one of the most famous archaeological monuments in the world, is usually a major reason for many visitors to come to Yogyakarta, and it is located forty-two kilometres to the north-west of the city. It is associated with two other Buddhist monuments of approximately the same period, Candi (meaning 'temple') Mendut and Candi Pawon, located very near to Borobudur itself. To get there by bus involves a change at Muntilan and also taking a garry or a pedicab from the bus station near Borobudur. It is certainly easier to hire a car for the visit if one is not going on a group tour.

Borobudur is a contracted expression meaning 'Mountain of accumulation of merits of the ten states of Bodhisattva', which clearly establishes it as a Buddhist monument of Mahayana persuasion. It is often held that it was conceived as a complete whole, and that the various levels of the monument represent different states of worldly being, moving from the temporal to the transcendental. This is theorizing after the event and on false premises. The archaeological evidence is that the monument was built in at least four different stages, that it was probably begun as a Hindu temple and transformed into a Buddhist place of worship after its second stage.

The first stage was begun about AD 775–80. The site was selected close to the meeting of two rivers, the Progo and the Elo, almost certainly to recall the place where the sacred rivers of Hinduism meet, the Ganges and the Jumna. The construction was on a foundation of volcanic tuff and consisted of the original foot and two galleries, laid out facing the four cardinal points in the shape of a mandala, with stairways linking the galleries in the centre of each side. There was another structure raised at the level

BOROBUDUR

of the third gallery which may well have been almost complete before being destroyed.

The second period of construction dates from around 790. The third and fourth galleries were added, the stairways were changed and doorways redesigned. A wide foot was added, almost certainly for technical reasons, to prevent the foundations giving way under the weight of the structure above. This covered the reliefs already completed in the first period (some of these have been uncovered in the south-west corner) and a circular structure begun on the top platform. Work then stopped for reasons unknown.

It was resumed about 810, and probably involved dismantling the circular structure already in place and building three circular terraces, the pierced stupas and the central stupa. Niches housing Buddha statues were added to the first gallery and further work was done on the foot.

The last two periods of construction, at about 820 and 840, involved no major changes but modified and improved the existing structure. Spaces between the niches on the first gallery were blocked and new reliefs added on the inside of the first gallery wall.

There was no mortar used and the stones locked into each other, giving considerable cohesion so long as the rain water did not erode the foundations. The carving on the galleries was done on the spot and some unfinished panels give an indication of the techniques used. The carvings were covered in two layers of stucco, which explains why details on the carvings are sometimes lacking; the stucco was almost certainly touched up once dry and was probably coloured as well.

The monument assumes the form of an enormous stupa raised on galleries which exist to allow for the Buddhist obligation to walk round a shrine in a clockwise direction; at the same time a pilgrim would be instructed through the sculptures found there. A pilgrim would have circled the monument in its final form nine times on reaching the top (Plate 8). The carving on the lower levels is both more interesting and better in quality. The upper level of the first gallery tells the story of the life of the Buddha from his birth to the sermon at Benares, and the lower levels depict episodes in the former lives of the historic Buddha. The second, third and fourth galleries illustrate Buddhist texts and the lives of Bodhisattvas. Some are not readily identifiable.

The first two terraces have respectively thirty-two and twenty-four stupas with lozenge-shaped openings and the third circular terrace has sixteen stupas with square openings. Each originally contained a Buddha statue in the posture of giving the law at the sermon at Benares. The central stupa contains two empty and sealed chambers.

Borobudur was probably abandoned as a shrine at the time of the shift in power to East Java. The monument was covered with vegetation when Raffles visited and described it in 1815. Serious archaeological work was not undertaken by the Dutch until much later in the century. Van Erp at the beginning of this undertook major restoration work on the circular terraces and the main stupa. But the original drainage system had become blocked and the foundations began to give way. An international project was launched in 1968 to 'Save Borobudur', involving the removal and relaying of the four lower galleries and cleaning the stonework. The monument was formally reopened by President Suharto in 1983.

A further project to create a park around the monument has still to be completed, but already the approach is far less worldly than it used to be, and vendors are forbidden within the temple grounds. The impressive tree-lined main approach is from the east, as with the original monument, but curves to avoid a village which had grown up near the base and which could not be removed.

From a distance, Borobudur does not give a particular impression of majesty or might. Its vast size and power are only appreciated as one goes around its galleries and rises up the steep stairways to the top. The view from the circular terraces across the valley and rice fields to the mountains and volcanoes is majestic, and from the top the importance of the monument can be appreciated.

Candi Mendut is three kilometres to the east of Borobudur, on the other side of both the Elo and Progo Rivers; it is actually the first temple one comes to on the road from Yogyakarta (Plate 9).

Like Borobudur, it was built in several different stages, and the earliest predates Borobudur itself by possibly twenty-five years, probably having been begun around AD 750. Also like Borobudur, it was probably first built as a Hindu temple. It underwent a transformation about 800, at more or less the same time that the massive third stage of Borobudur was undertaken, and Mendut's third stage was begun about 830.

In appearance it is, however, quite unlike Borobudur. It is in the form of a square cella (the inner part of the temple housing the sacred images), with a portico serving as an entrance to the shrine, the whole raised on a massive foundation. There are three statues inside the cella; the central figure shows the Buddha seated in the European fashion teaching the law (the symbolic gesture, or mudra, known as *dharmachakra*), with Bodhisattvas on either side. Avolekitesvara is to the left of the Buddha and the statue to the right is usually identified with Manjusri. All three are raised on a plinth, at the bottom of which is the Wheel of Law with two gazelles, a reference to the gazelle park in Benares where the Buddha preached the law for the first time. The quality of the carving is excellent and the huge statue of the Buddha one of the finest in existence. Originally the face of this statue was lit by the evening sun coming through a window above the entrance, but subsequent transformations of the entrance portico caused the opening through which the light came to be blocked.

The sculpture on the outside in relief is no less arresting. The wall panels are decorated with Bodhisattvas protected by parasols. The approach stairway has carvings illustrating stories from the Jataka tales. Inside the entrance portico itself there are also carved panels.

Candi Pawon is very small, located half-way between Candi Mendut and Borobudur, and slightly off the main road to the left. It has a square plan with, like Mendut, a portico, though the proportions are much reduced. It contained a single statue. The whole building was heavily restored in the early part of the twentieth century and the present roofline may be far from the original. The temple has local associations with Kuvera, the Hindu deity associated with wealth, but this seems unlikely, given its obvious link with the two Buddhist temples nearby, unless, like them, it was also originally a Hindu temple. However, there is no evidence of any changes of plan to Pawon and there are no indications of its date of foundation, though it seems likely to have been about 800. The name, meaning 'kitchen' in Javanese, is no help; most of the current names of temples were taken from nearby villages or given much later in quite recent times. The name Borobudur is an exception, being original.

## The Prambanan Group

*Prambanan, Kalasan, Sambisari, Sari, Sewu, Plaosan, Ratu Boko*

The village of Prambanan lies seventeen kilometres east of Yogyakarta on the main road to Solo. The principal temple, given its size and importance, is Candi Prambanan, sometimes referred to as Candi Loro Jonggrang, and is clearly visible to the left of the road (Plate 10). To refer to the temples nearby as a group is perhaps confusing, since they are only grouped by geographical location and not by date or religious affiliation. Because the temples are not particularly close to each other, transport between the sites is necessary; while one can take a bus to Prambanan, it is not possible to walk to all the temples. The use of a car or the back seat of a motor bike is recommended. The seven major sites (and there are, in addition, several minor ones) comprising the group are perhaps less well-known internationally than Borobudur, but are no less important and in some ways more satisfying visually. The temples in question are, with dates, sometimes approximate:

| *Buddhist* | | *Hindu* | |
|---|---|---|---|
| Kalasan | 778/790/810 | Sambisari | 835 |
| Sewu | 780/792/810 | Prambanan | 856 |
| Sari | 835 | | |
| Plaosan | 835 | | |

Ratu Boko ninth century
(possibly both)

The oldest building in the Prambanan group, and the first encountered on coming from Yogyakarta, is Candi Kalasan, to the

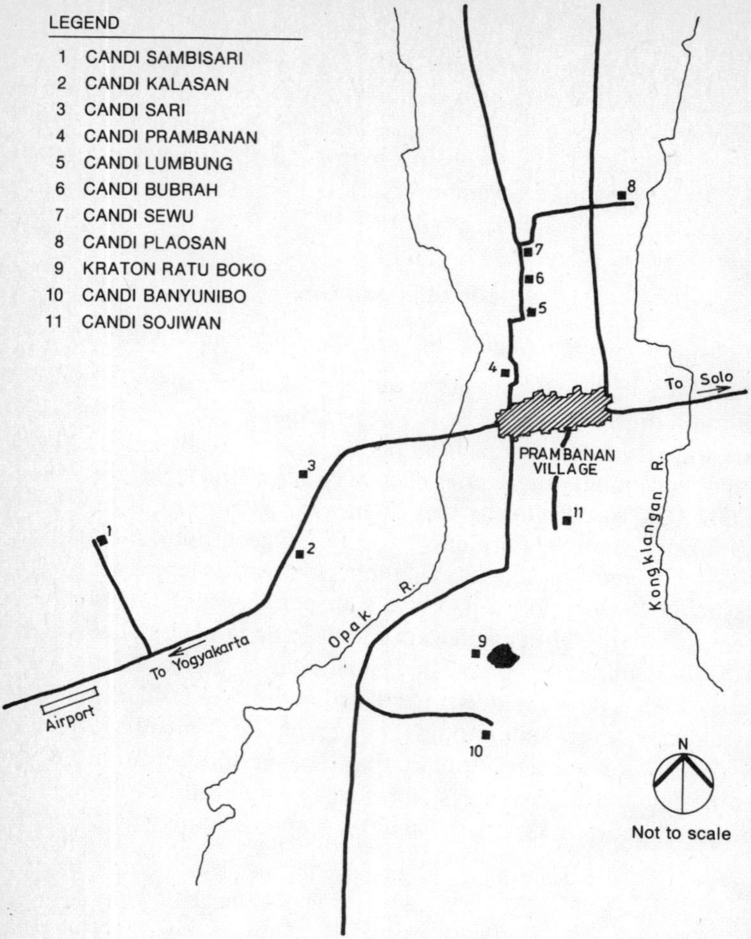


PRAMBANAN GROUP

right of the road just after the airport in a village of the same name. A stele to commemorate the construction of the Buddhist temple and which can be dated to 778 dedicated the building to the Goddess Tara. The buildings which can be seen today (and where archaeological work is currently being undertaken), are the result of the third transformation of the temple, though the major change, from a square plan to a cruciform plan, was effected in the second construction. There are no statues remaining, though the central cella would have had a Buddha statue against the west wall on a plinth, in a similar fashion to that found today at Mendut. There are three niches for statues on the access corridor to the central cella, and the shrines around this on the outside would also have had three statues, the central one seated. The decoration around the outside niches is elaborate and impressive. Above the doorways can be seen Buddhas and also Kalas, and high up forming a kind of pediment around the central tower in the form of an octagon covering the central cella, are small niches with seated Jinas. All the decoration was originally covered in stucco and probably coloured as well. In the grounds are several *makara*, mythical beasts, originally forming the lower ends of retaining walls for stairways (Plate 11).

Closer to Yogyakarta, one and a half kilometres after the airport turn-off and to the left, is a rough road going to the village of Sambisari. The excavation has only been recently completed of the Hindu temple found here (Plate 12). It was covered with volcanic ash about the end of the fourteenth century and was, because of this, very well preserved. A linga placed on an elaborately carved plinth showed that the temple was dedicated to Siva. There were three small terraces in front of the main temple and a wall going all round the precinct. The actual proportions of the temple are small, and it has sunk well below the present level of the land.

Further along the Yogya-Solo road, and only a short way beyond Candi Kalasan, is a track to the left leading to Candi Sari (Plate 13). The plan of this is unusual (and to be replicated in Candi Plaosan), being a double-storey building, facing east, with three cellas on each level, one above the other. The second storey was reached by a wooden staircase and the first step in stone is still there. It is sometimes erroneously maintained that the second storey was a priests' dormitory, but buildings of stone were reserved for the gods (mortals occupied wooden dwellings) and no Buddhist would ever place himself above a representation of the Buddha. None of the statues, most probably three in each cella, has survived in place. The entrance to the building and the plinth around it are defective. The quality of the carving on the outer walls is remarkably good, showing, among others, Bodhisattvas with musical instruments. The restored roofline bears a close resemblance to the decoration topping the walls of the first gallery at Borobudur.

After crossing the Opak River just before the entrance to the village of Prambanan, you take a side road to the left to reach the main temple.

Prambanan is an enormous complex with three different enclosures, though almost none of the buildings in the outermost remains. There is one surviving entrance, now incorporated in the buildings for the modern theatre for the Ramayana ballet. The second enclosure contains the ruins of 224 shrines known as the Candi Perwara arranged in three concentric rows with their doorways facing the different cardinal points. The inner enclosure has the central shrine, dedicated to Siva, with one to Visnu to the north and another to Brahma to the south (that to Visnu is currently being restored). Opposite are the ruins of three temples dedicated to the mounts of the Hindu gods, the bull Nandi for Siva (with its splendid statue still in position), the mythical bird Garuda for Visnu, and Hamsa, the sacred goose, for Brahma. There are two other small shrines and nine chapels marking the place of the stones used to set out the complex.

The main shrine, like the rest, when Raffles visited it (and he seems to have been the first person in modern times to have taken an interest in the monuments of Java) was covered in vegetation and 'is at present a mass of ruin'; he noted that the Javanese, though nominally Muslim, continued to pay honour to the Hindu statues there. The shrine has now been restored and is magnificent. The central cella is occupied, not by a linga as was usually the case in Sivaite temples, but by an enormous statue of the four-armed god in human form, facing east; the venerable sage Agastya is to his right, behind him is his son in elephant form Ganesha, and his wife Durga is to his left. She is the 'slender maiden' of innumerable local legends, and offerings of flowers for domestic fertility can always be found at her feet.

This central shrine is raised on a high foundation, with excellent carving on the outside, and on the inside of the balustrade of the walkway linking the four chapels are carved panels showing scenes from a Javanese variant of the Ramayana. Not all the scenes can be readily identified, but those showing Sita being taken by Rawana (and in the same panel the bird Jatayu giving Sita's ring to Rama to show where his wife is) and Hanuman being made prisoner and escaping by setting fire with his burning tail to Rawana's palace, are easily recognizable. The story continues on the balustrade of the temple to Brahma to the south and the carvings on the temple to Visnu illustrate stories from the life of Krisna, one of the incarnations of Visnu.

The restored central shrine to Siva at Prambanan is a monument of major importance and is in many ways more immediately impressive than Borobudur, because its verticality and not its horizontality is stressed. The architects strove to create the impression of a Mount Meru reaching into the heavens to the glory of the forms of the gods beneath, and undoubtedly succeeded. The presence of so many other shrines in the complex adds to the effect of the site.

To the north of Candi Prambanan are two small temples, Candi Lumbung and then Candi Bubrah, both fairly ruined. Beyond these lies the huge Buddhist complex of Candi Sewu (meaning 'one thousand shrines').

Candi Sewu is two kilometres to the north of Candi Prambanan and, like Candi Kalasan, was constructed in three different periods. The first probably dated from the same time as Kalasan itself, or a little later. The temple complex is guarded by raksasa, or giants, armed with clubs or staves (Plate 14), and there are 240 small shrines in the second precinct. This large number undoubtedly accounts for the name of the temple. Each shrine has the same plan, with a square cella and a small portico in front. The temple underwent a major transformation in AD 792 and this particularly affected the central sanctuary, which seems to have been adapted to a different religious rite. All the statues were renewed at this period, and changes were made to the 240 small shrines as well, incorporating representations of the five Jinas.

One of the small shrines has been restored and major restoration work is in progress on the central shrine. Though Candi Sewu in some ways looks like a builder's yard, and probably did so throughout much of its existence, it is nonetheless an impressive site, and has all the fascination that ruins exert.

Some two kilometres to the east of Candi Sewu is Candi Plaosan (Plate 15). Like Candi Sari, the central shrines (there are two main ones, the southern being the larger) were built on two levels, and in both the stone step forming the base for the wooden staircase to the upper floor is in place. Unlike Candi Sari, both sanctuaries face west. The statues on the lower levels would have been similar to Candi Mendut, with a Buddha statue in the centre and a Bodhisattva on either side, though on the upper storey there may only have been one statue. The stone Bodhisattvas are still in place, but the central Buddha statues, probably of metal, have disappeared. There are reliefs on the walls inside the main sanctuaries, possibly showing donors, certainly in attitudes of devotion towards the central statues. The outsides of these shrines were carved in a similar fashion to that found at Candi Mendut; the carvings on the rear wall, facing east, of the restored southern sanctuary are as impressive, and so is the roofline. Each of the main sanctuaries was guarded by a pair of raksasa, as at Candi Sewu.

These two central shrines have a wall around them and outside this are 174 smaller constructions, consisting of stupas and shrines, arranged in three concentric rows, the gifts of officials. As with Candi Sewu, restoration work is also in progress here, and a great deal remains yet to be uncovered; to the north of the main shrines is an area where excavation has barely begun. To the south of the present entrance to the complex, on the other side of the minor road and a little lost in a bamboo grove, is another small shrine associated with this temple.

Candi Plaosan is the last great Buddhist monument built in

Central Java and when one considers that Prambanan, Sari, the last two transformations of Kalasan and Sewu, Lumbung, Sambisari and several minor temples which have largely disappeared were all built in close proximity in the space of about fifty years (not to mention the major transformations of Borobudur), it seems likely that the building effort must have exhausted the Javanese in the same way that the huge constructions at Angkor did the Khmer.

Seen today, Plaosan and Sewu in particular, with their views across peaceful rice fields towards smoking Merapi to the north, are powerful testimonies to the glorious past, to which the present, with its humble villages and farmers, forms a distinct contrast.

Kraton Ratu Boko is located high on a plateau overlooking the Prambanan plain, to the south of the village of Prambanan. To reach it, turn south instead of returning on the Yogya road, before crossing the Opak River. There is a signpost on the road indicating, not very clearly, the track leading up the hill to the site. After a short and fairly stiff climb (keep to the left at the top), the path levels off and there is a splendid view over the plain. There is a double ceremonial porch, which has been restored, and slightly to the north of this the foundations of two buildings, one probably a temple. Many visitors think this is all there is to see. However, turning slightly south and then east, passing through the paths of a very poor village, one comes to the major part of the ruins, now extensively restored. Here are stone foundations, precinct walls, central wells indicating shrines of some kind, most probably Hindu, and stone bases for wooden pillars clearly intended for pendopo structures. Beyond this confusion of foundations and walls, and at a lower level, is a series of pools which may or may not have served as bathing pools. Although the area is known as a kraton, or palace, and though legends abound about its occupants, there is no firm evidence that it was ever such. At present, very little is known for certain about the purpose of these buildings.

Raffles quoted a Captain Baker in *The History of Java* who noted:

> In the enclosures to several of the villages (which are here frequently walled in) are discovered large stones, some representing gorgon heads, others beautifully executed in relief, which had formed the frizes [*sic*] and

cornices of temples, all regularly cut so as to be morticed together, but now heaped one upon the other in the utmost confusion and disorder.

This can be clearly seen in the village leading to the raised platforms and sunken pools of Ratu Boko; there are many carved stones to be found here forming rough walls to house compounds.

This by no means completes all the temples in the Prambanan area, only the main monuments. A little further along the road leading to Kraton Ratu Boko, beyond the hill with the path going up to the site, is Candi Banyunibo, another ninth century Buddhist temple in the middle of rice fields to the left of the road. It has been indifferently restored but has interesting carving, especially above the entrance. It is very difficult to reach in the rainy season, however. Remains of another Buddhist temple, Candi Sojiwan, lie some two kilometres to the south of Prambanan village (the side road is marked at the eastern end of the village). Candi Sewu was originally surrounded by four temples at each of the cardinal points; Candi Bubrah to the south has already been mentioned, and the remains of one of the others can still be seen in the village facing Candi Sewu to the east.

## Dieng

The oldest temples in Java are to be found on the Dieng Plateau, 135 kilometres north-west of Yogyakarta. If going there by bus, go to Wonosobo and then by a communal taxi or minibus. If going by car, care is needed on the final twenty-five kilometres from Wonosobo up to the plateau, 2 000 metres above sea-level; the view is spectacular along a narrow bending road not always open to traffic in both directions. If staying overnight (there are fairly basic hotels), warm clothing is necessary. Horses can be hired to visit the sites.

Dieng occupies a collapsed volcano rim and the ground is marshy. There are geysers, sulphur fumes and small lakes. Originally the area was drained and some of the underground culverts remain. The site was occupied from the end of the seventh century to the thirteenth century and there are traces of residential buildings. Eight of probably three times that number of temples can be seen today; they all bear the names of the Pandawa heroes in the Mahabarata, but these are certainly not the original names.

The earliest buildings are Candi Arjuna, Semar, Srikandi and Gatokaca, and date from the late seventh century to about AD 730. The style is very sober. The second group, comprising Candi Puntadeva, Sembrodo, Bima and Dwarawati, were built between 730 and 780 approximately, and these are more elaborate. Five of the candi, Semar, Arjuna, Srikandi, Puntadewa, and Sembrodo are all located quite near each other, between the remains of a supposed palace and a monastery. The rest are scattered on the plateau. All were given over to the cult of Siva, who would have been represented by a linga in the shrines.

Perhaps the most impressive temple is Candi Bima, built on a quite different model to the others which are mostly fairly simple, with a central cella entered through a small portico and with little decoration (Plate 16). Bima, however, located not far from the 'coloured lake', is more complex and has no parallel in Java. It has a fascinating tiered roofline filled with small figures.

The monuments of Dieng are transitional between Indian models and more evolved Javanese types; they are of relatively modest proportions and nowhere near the scale of the grandiose temples of a century later. The site is unusual; in this rarefied atmosphere, the gods, one feels, are close at hand.

## Gedung Songo

These 'nine buildings', as the name of the site translates from Javanese, are all Hindu temples probably built between AD 730 and 780, located on the southern side of Mount Ungaran, some eighty kilometres to the north of Yogyakarta. Using public transport, the bus goes to Ambarrawa and there is a twelve-kilometre andong ride from there through Bandungan to a point three kilometres beyond, where the road starts to the temple sites. If going by car, vehicles can be parked at the top of the road. It takes about two hours to climb to the major shrines, but the effort is well worth it.

The temples are scattered over the summits of six eminences on the mountainside and the views over the plain and distant volcanoes of Merapi, Merbabu and as far as Lawu are superb: the site is one of the most impressive in Java. The main shrine in each temple was dedicated to Siva; not all the others can be identified. The buildings are divided into five groups and the problem of names is overcome by referring to them simply by numbers. The first is a single temple and a path leads from this to the others. Group 2 and Group 3 are to the north of this; the path turns west across a small ravine with a sulphur spring. Group 4, with a temple to Visnu, is to the right of the track at the top of the buff. At the edge of a sheer drop, are the several shrines of Group 5, to the left at the top of the buff.

The original plan of the temples was modest and fairly uniform, with a simple central cella opening onto a narrow corridor with a portico at the head of a short stairway rising above the raised foundation. Some of the temples have been restored, some are in ruins.

Gedung Songo, by virtue of its dramatic and rather isolated site, across which the wind whistles unforgettably, as much as for its archaeological remains, is well worth a visit and is in some ways more rewarding than the better-known Dieng Plateau.

# 7
# Around Yogyakarta

## The Grave Sites

*Kota Gede and Imogiri*

KOTA GEDE is now a suburb of Yogya, five kilometres to the south-east of the city. It was, however, the capital of the Mataram kingdom at the end of the sixteenth century. Virtually nothing remains of its former importance except for the tombs of two rulers of Mataram and their relatives, which are all found in a single building not far from the market.

The rulers were Senopati (reigned *circa* 1584–1601), and his son Krapyak (?1601–13), who was the father of Sultan Agung. The expansion of Mataram, which began with Senopati, found its apotheosis in his grandson, who is buried at Imogiri. People go to visit the tombs at Kota Gede, in traditional Javanese dress, which is obligatory, to kneel on the stone floor, principally by the tomb of Senopati, and murmur prayers to the spirits of the departed kings.

A path by two large wringin, or banyan, trees leads into a courtyard from where one can take a path to an old mosque or visit numerous other smaller courtyards. The steps leading to Senopati's tomb in the cemetery are worn and strewn with flower petals. There are walled gardens around. By the buildings containing the tombs is a pool with huge fish and turtles and the approach has an air of decaying grandeur.

Kota Gede is perhaps better known as a silverware centre. Central Javanese silver is elaborately wrought and not to everyone's taste. With the rise in recent years of the price of silver, the craft has somewhat diminished in importance, but a number of establishments still continue to give employment to traditional silversmiths.

Otherwise, Kota Gede is without interest and boasts the worst roads in Yogya.

Imogiri is some twenty kilometres south of Yogya, on the eastern side. The road passes through a lush countryside of rice paddies, orchards and villages. Trees overhang the narrow highway, streams and irrigation channels flow fast, old ladies pass constantly to and fro, armed with sickles, with huge wicker baskets on their backs held by slendang, or scarves, and wearing small conical hats. To reach the royal tombs at Imogiri, one passes through the village to the base of a hill where the road stops. The site is one of the hills forming part of the low range known as Gunung Sewu ('a thousand hills').

The tombs are located on top of the hill and are reached by climbing a broad and not particularly straight cement stairway. The great Mataram king, Sultan Agung (reigned 1613–46), is buried here, together with his descendants. During the period 1614–22 he established his court at Karta, some five kilometres south of Kota Gede. He prepared his own grave site at Imogiri, another five kilometres south of Karta, the year before he died, after a tumultuous reign during which he only rested in the last years, from 1641 when he received from Mecca the authorization to style himself Sultan. (He started as a mere Pangeran, lord, and then became Panembahan, honoured lord; after conquering Madura in 1624 he took the imperial title of Susuhunan.) The actual burial chamber of the Sultan has been compared to a small bedroom with white hangings on the walls and the black tomb in the centre. Incense and flower petals abound and Javanese dress is required to enter the mausoleum.

After the split in the remnant of the kingdom of Mataram in 1755, both the Susuhunans of Solo and the Sultans of Yogya were also buried at this site, the Hadiningrat line, being the senior, to the right of Sultan Agung facing the sea (that is, to the left at the top of the stairway), and the Mangkubumi descendants to the left of the Sultan.

At the top of the long stairway is a split Balinese style gateway, with beyond that a closed Balinese style gate, in brick and stonework. The split gateway is very thick, perhaps for technical reasons, and relatively low. The motifs on the closed gate are similar to those found at the fourteenth-century temple of Penataran in

East Java. On two platforms to the right and left of the open space at the top of the stairway are small pendopo for the guardians of the place. Two large martaban jars are found here. They are sixteenth-century Thai wares and have been given male and female characteristics by the Javanese, and the names Kyai Mendung and Nyai Sijem (the latter meaning 'gift of Siam').

The view from the top towards the sea, some fifteen kilometres away, is not particularly remarkable, over the low hills with a sparse vegetation of teak trees. It is said that Sultan Agung used to sit here to commune with the Goddess of the Southern Ocean, Ratu Loro Kidul.

On special occasions in the Javanese traditional calendar, Imogiri is worth a visit for the atmosphere of hushed reverence of the crowds paying respects to the tombs, but on ordinary days the place is somewhat characterless.

## The Coast

*Parangtritis, Samas, Baron*

The coast to the south of Yogya is extremely inhospitable. There is no land between Java and Antartica and the depths of the Java Trench are not far off. The swell off the coast is considerable and there is a very treacherous undertow along the whole of the coast, which is without natural harbours except, to the east a good way from Yogya, at Pacitan and, even further in the other direction, to the west at Cilacap. Fishing is therefore scarcely practised on this coast, and the excellent fish obtainable in the Yogya markets come from the north coast, at Semarang, some 120 kilometres away.

The southern ocean is considered to be the domain of Goddess Ratu Loro Kidul, who lives beneath its waves and is in communion with the founder of the Mataram kingdom, Senopati, and his descendants, including all the present line of sultans. Annual offerings are made to the Goddess, including the nail parings and hair clippings of the reigning monarch.

Ricklefs, in his *History of Modern Indonesia*, states:

According to Mataram tradition, she was a princess of Pajajaran who had been driven from the court when she refused a marriage arranged by her father. He laid upon her a curse: she was made queen of the spirits with her palace beneath the waters of the Indian Ocean, and would only become a normal woman again on the Day of Judgement. She is supposed to have encountered Raden Susuruh, whom *Babad Tanah Jawi* texts inaccurately depict as the founder of Majapahit, and to have told him that she would serve him and all his descendants who ruled the area of Mataram. After Senopati spent three days with her in her underwater palace, the Goddess promised him the support of her spirit army. . . .

Ratu Loro Kidul is said to particularly like the colour green and to snatch people wearing clothes of this colour to her lair beneath the waves. She is also said to lure young men who think they are stronger than her to their deaths by enticing them to swim in the swirling waters of her green sea. Whatever the strength of Loro Kidul and of legend, her waters are not to be trusted and numerous people have been carried out to sea who have tried to swim here. This coast is extremely dangerous and no swimming should be attempted.

Parangtritis is the centre of the cult of Loro Kidul, more particularly at Parangkusumo, a little along the shore from Parangtritis, where offerings are made to the Goddess at the annual Labuhan ceremony. The fine grey sand beach is overhung by cliffs with caves in which hermits live. There are guest-houses, freshwater pools for swimming, and warung, or wayside stalls, shaded by coconut trees. Magic mushrooms are also a speciality of Parangtritis.

Although it may seem a little perverse, Parangtritis, which is twenty-eight kilometres from Yogya, is not best reached from the city by taking Jalan Parangtritis. This comes to an end at Kretek by the Opak River, which in the rainy season, usually from October to March, is in flood and only small motor bikes and bicycles can be taken over on a ramshackle bamboo ferry for the last five kilometres to the beach. Buses and most other vehicles now approach Parangtritis by taking the Imogiri road, for there is a bridge over the Opak along this and no hazards of fords or floods.

For a wind-swept dangerous coast, the special reverence with which Parangtritis is held by Javanese is at first difficult to understand, but it combines in a very Javanese fashion a heady mixture of primordial divinity (the goddess with whom the ruler is in special communion and to whom he makes offerings), mysticism (kebatinan, or Javanese meditation, is much practised on this beach), magic, nationalism (the nineteenth century hero Diponegoro hid in one of the caves here) and a touch of the erotic (Loro Kidul is also a syren who lures young men to her lair).

The beach of Samas is nearer Yogya, being twenty-five kilometres away, and is reached by taking Jalan Bantul and passing through this town, a centre of production of sugar-cane and vegetables. The beach is treeless, windswept, and uninviting, as well as dangerous. There is accommodation for the hardy.

Baron is a small sheltered cove, some sixty kilometres south-east of Yogya. The road, in very poor condition, goes through the poorest part of Central Java after Wonosari, in the centre of the barren limestone hills of Gunung Kidul. There is very little topsoil, and what there is is washed out of the hollows in the rocks by the rain. Old women go to the bottom of the gullies with baskets to scoop the earth up and replace it in the hollows. Only cassava and maize manage to grow here, and not much of either. The trees are stunted. The poverty is grinding.

Bathing is possible at Baron because the tiny bay is almost entirely enclosed from the sea; however, swimming must be confined to well within the bay, especially when the tide is ebbing. During the rainy season, a freshwater stream turns the waters of the cove brown with mud, but in the dry season they are clear. There are a few warung here, but the resources of the area being limited, little should be expected.

Kukup is a tiny beach of white sand a little beyond Baron to the east. The track leading there is clearly marked. The beach is rather exposed and bathing is extremely dangerous here. The currents and undertow are very strong.

The southern coast of Java, as can be seen from these remarks, is not a swimmer's paradise and in no way does it conjure up the standard image of a tropical idyll.

## The Mountains

*Merapi, Kaliurang, Salatiga*

Mount Merapi (meaning 'giving fire') is a highly active volcano about thirty kilometres to the north of Yogya and, except in the rainy season, is constantly seen brooding over the city. It emits gas, boulders, ash and hot mud fairly regularly, and occasionally produces spectacular lava flows. There are six observation posts high up on its slopes to record its daily activity and give warning of eruptions to local inhabitants.

The main damage caused by the volcano is less in the lava flows than in the washing down of rocks, sand, mud and volcanic matter by heavy rains. These block the streams and rivers, cause floods and adversely affect farmland, dams and irrigation systems, roads and bridges. On the road to Magelang, near Muntilan, can be seen some of the effects of recent eruptions; bridges have had to be rebuilt, and the fields and river beds are strewn with rocks and boulders.

To reach the top of the volcano, go to Kartosuro on the Magelang road, and then take the direction of Boyolali, stopping at Selo, where there is an observation post and guides can be hired. It is essential to enquire first about the current state of volcanic activity at the observation post. Most people start the climb to the summit just before dawn, returning about midday. The view from the top, almost 3 000 metres above sea level, is dramatic.

For the less adventurous, Kaliurang, on the slopes of Merapi, about twenty-five kilometres north of Yogya, is almost as rewarding and a good deal less exhausting to visit. It is a mountain resort very popular with young people on weekends, for bungalows can be hired and the cool climate makes a pleasant change from the heat of the plains. There are a couple of swimming-pools and an attractive waterfall, tennis-courts, guest-houses and a hotel. The place has pleasant gardens and the outskirts are wooded.

A short walk of less than an hour brings one to the Plawangan lookout, with a fine view of the volcano and the broad plain looking towards the southern ocean, edged by ranges of hills to the right and left.

Salatiga is a pleasant cool mountain town on the road from Solo to Semarang. It can be reached by continuing along the road from Kartosuro to Boyolali, running between the twin volcanoes of Merapi and Merbabu, and then turning left. It is a fruit and vegetable growing area and boasts a large Christian university, Satya Wacana, but is of no special interest to tourists. Javanese like to go to swim there in the cool of the mountains on weekends. It is relatively near Mount Ungaren and the temple sites of Gedung Songo, but these are more conveniently reached from Yogya via Magelang. Nearby Salatiga is the Rawapening lake with unusual floating rice paddies.

## Solo

Solo, also known as Surakarta, and often written Sala, is sixty-three kilometres to the north-east of Yogyakarta, and a large commercial and cultural centre in its own right. It is much less popular with overseas visitors than Yogya, though it has residual courts, like Yogya itself, and has a thriving, if less artisanal, batik industry.

Although nowhere near so gracious and relaxed as Yogya (a statement its partisans will dispute), it maintained and developed over the years, especially in the nineteenth century, its own cultural traditions, particularly in the dance, as a separate seat of power of the split Mataram kingdom. Solo is worth visiting on account of its courts. Furthermore, it is on the way to Mount Lawu, which has two temples on its slopes, Candi Ceto and Candi Sukuh, the second of which, though one of the latest of the Central Javanese classical monuments, is entirely original and unexpected.

Solo was the capital of Mataram from 1746 until the kingdom was partitioned, with the founding of Yogya, in 1755. The capital had previously been twelve kilometres to the west, at Kartasura, which itself had been founded in 1680 by Amangkurat II. A descendant, Paku Buwono II, reigned disastrously from 1726–49 in a period of constant warfare involving the VOC, the Balinese ruler of East Java, Surapati, the ruler of Madura, Cakraningrat IV (who occupied Kartasura for a time in 1742), and numerous others. Paku Buwono II decided to abandon Kartasura in 1743, but did

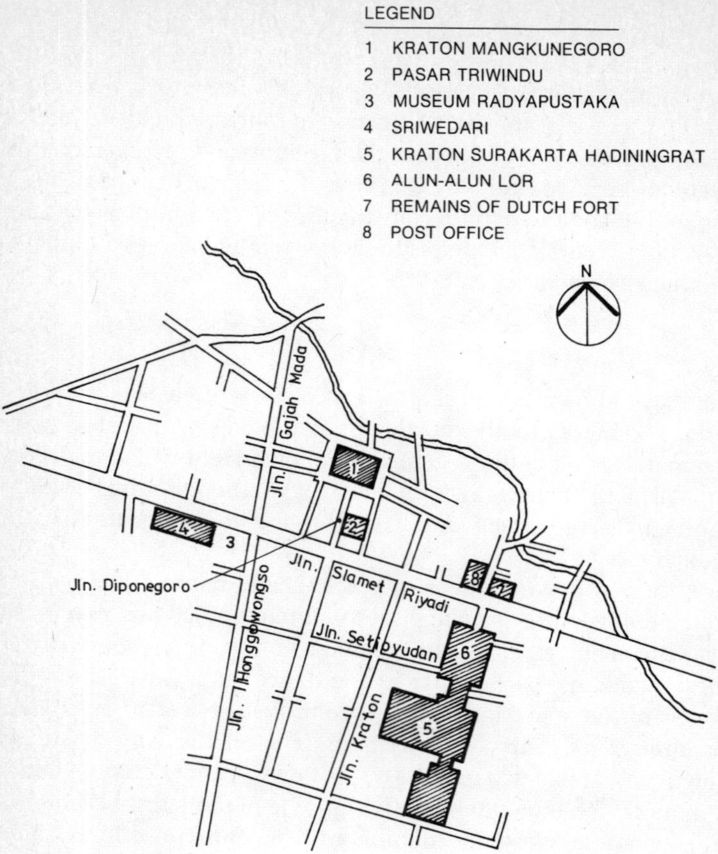


SOLO (SURAKARTA)

not live long to profit by his new capital. The ruins of Kartasura, which are few and of little importance, can with difficulty be found and seen in the village of the same name.

Paku Buwono III, after the split of the kingdom, remained in Solo as head of the Hadiningrat line, while his uncle Hamengku Buwono I established his capital at Yogyakarta. Later, the unfortunate Susuhunan, often translated by contemporaries into English as 'Emperor', had his lands further reduced in size with the establishment of the principality of the related Mangkunegoro family in 1757, made permanent by the Dutch in 1792 when it was decided the Mangkunegoro heirs would inherit the title and lands that went with it. Yogya in 1813 was to suffer the same fate with the creation by Raffles of the Paku Alaman.

Neither of the Solo princely families managed to adjust to the post-war situation in Indonesia and were seen to be too close to the Dutch. Paku Buwono XII and the subsidiary prince Mangkunegoro VIII, both of whose reigns started in 1944, had them cut short on 1 June 1946, after which date their prerogatives were confined to the walls of their courts. Unlike Yogya, their territories were absorbed into the province of Central Java. Their kraton remain, of course, and particularly the Mangkunegoro kraton, part of which serves as a hotel now, should not be missed.

On the main road, Jalan Slamat Riyadi, arriving from Yogya, on the right, is a small museum, the Museum Radya Pustaka. This has in particular a number of artefacts and statues from the eighth to the tenth centuries, as well as a collection of gamelan instruments, wayang puppets, royal relics, models, and books. Though not very well-displayed, it is worth a visit.

Next to this museum is Sriwedari, the home of one of the most famous Javanese classical dancing troupes, which specializes in wayang wong, Javanese dance-dramas. Sriwedari is also an amusement park.

Further towards the centre, turning left at Jalan Diponegoro, is, at the end of the street, the Kraton Mangkunegoro, still occupied by the princely family who built it. It is one of the finest of all Javanese palaces, centring round a particularly impressive main pendopo, with private chambers behind this. There is a small collection of artefacts kept here and the antique gold jewellery of the Majapahit and Mataram periods is exceptional.

The palace was extensively restored from 1917–20 and again from 1937–42 by Thomas Karsten, a Dutch architect working in Central Java (he was also responsible for the building of the Sono Budoyo Museum in Yogya, on the Alun-Alun Lor). The palace as a whole is in a remarkable state of preservation, and says something of the ability of the Mangkunegoro family, notably since 1965, to adapt to more recent circumstances. One is also allowed into a number of rooms in the private apartments in the palace, which are both tasteful and revealing. The gamelan here is excellent and occasionally dance performances in the main pendopo are held.

In the same street, Jalan Diponegoro, which leads to the Kraton Mangkunegoro, very shortly after turning from the main road from Yogya, is, on the right, the very picturesque Pasar Triwindu. This is the only authentic flea-market in Central Java. The days when old Dutch oil lamps, Chinese porcelain and Mataram kris could be had for a song have long since gone; most of the stall holders know their prices all too well. But one can still find good pieces here at prices often well below what one would have to pay in an antique shop, and in addition there is the fun and fury of rummaging through an enormous quantity of plain junk as well as good antiques. The rear part of the market specializes in spare parts for car engines and electrical machinery. It is well worth a visit. Bargaining is essential.

The Kraton Surakarta Hadiningrat is in the centre of the town, south of the eastern end of Jalan Slamet Riyadi leading from Yogya. Like its counterpart, it has an Alun-Alun Lor with two central wringin trees and the main part of the kraton constitutes a series of enclosed courtyards. The palace as a whole was, especially in comparison to the Mangkunegoro palace, rather ill-arranged and dilapidated, and recently suffered from a serious fire. In an inner courtyard is a strangely shaped tower which is where the 'Emperor' was supposed to communicate with the Goddess of the Southern Ocean, Ratu Loro Kidul (a privilege shared by the Sultans of Yogyakarta). The kraton has a museum section, the Museum Radyapustaka, with Hindu-Buddhist bronzes, Chinese porcelain, Javanese puppets and royal pusakas, or relics. There are some figureheads from royal barges, used by the rulers to travel on the Solo River, one of the longest in Java, which flows into the sea near Surabaya. Above all, the museum has some marvellously decaying coaches, in which the Susuhunan used to be transported on ceremonial occasions. The oldest is a gift from the VOC to the 'Emperor' when he still resided at Kartasura.

Just beyond the western gate of the Alun-Alun Lor is the enormous Pasar Klewer, devoted exclusively to cloth, mostly locally-produced batik, and more than filling the two floors of the market. It is both wholesale and retail and not a little confusing in its huge variety. It is well worth visiting, though one tends to be overwhelmed by the diversity and quantity of the goods on display. The main market, the Pasar Besar, is very similar to that in Yogya except that it is also on two floors; it is quite near the Pasar Klewer.

Candi Sukuh and Candi Ceto are located to the east of Solo, on the western slopes of the volcano Mount Lawu. To reach the road leading to them, go through the city following the main road to Surabaya, and not long after crossing over the Solo River, where the main road veers to the left, the minor road leading to Tawangmangu on Mount Lawu continues straight ahead. This climbs gradually at first, and then very steeply indeed before the branch leading to Candi Sukuh ends.

Candi Sukuh is a remarkable construction, quite unlike any other temple in Java. With Candi Ceto, it is in fact the last of all the temples of the Hindu-Buddhist period and appears to date from about 1430. However, its extremely crude carving and extraordinarily close resemblance to central American Mayan pyramid temples made early commentators think it was one of the earliest Javanese constructions. Raffles, in his *History of Java*, includes several pages by a Dr Horsfield who visited the monument in May 1815 and comments that it had only been known about to Europeans a short time before thanks to the British Resident at the court of Surakarta:

> The principal structure is a truncated pyramid, situated on the most elevated of three successive terraces. The ruins of two obelisks, having the form of a section of a pyramid, are also observable in the vicinity of the principal building, and on each side of the western front appear several piles of ruinous buildings and sculpture. . . .
>
> The approach is from the west, through three porches or gateways, of which the outermost alone is now standing . . . in relief in the centre of the flooring under the porch is a representation of the male and female pudenda. . . .

The temple is built of blocks of laterite, and the central pyramid, which has a stairway leading to the top from the inside, has a level open platform with a commanding view over the countryside. This building is virtually without decoration. All around the site are strewn carved panels, mostly apparently dedicated to the cult of Bima and Arjuna, wayang heroes, which indicates to what an extent traditional Hindu-Buddhism had decayed by the time the temple was built. This isolated temple is mysterious and entirely unexpected.

Candi Ceto, which dates from about 1470, is further up the mountain and located slightly north of Candi Sukuh. It is less interesting than Sukuh and in recent times has been very extensively restored, to such a degree that its authenticity is somewhat doubtful. It consists of seven stepped terraces on which wooden pavilions have been placed. There are sparse remains of indifferent sculpture, including a phallus pointing towards the tip of a triangular stone, the whole decorated with lizards, toads and crabs. Again, it would appear that it was dedicated to the cult of Bima and is an indication that traditional religious inspiration had waned to such a degree that the advance of Islam, newly arrived in Central Java, was unlikely to find strong opposition, particularly given the syncretic nature of religious belief in Java. The approach road is sensational.

## Semarang

Semarang is not normally on the average tourist's itinerary during a visit to Yogya, but it is the largest city in Central Java, the chief city of that province, and a place that figures large in the history of both Yogya and Surakarta by virtue of it being the seat of the Dutch Governor of the North-East Provinces.

It is 120 kilometres north of Yogya on the attractive road through Magelang, but conforms to the general fact that the larger the city in Indonesia the less attractive it is. Semarang is a big, noisy, dirty city of more than a million inhabitants, squashed between the silted shoreline and the high mountains overlooking the narrow coastal plain and, on that account, extremely hot. This said, if time allows, Semarang is not without interest.

Semarang was ceded to the VOC by Amangkurat II of Kartasura in agreements of 1677 and 1678 as a direct cession in part-payment of his debts, run up for military services in the ceaseless wars of his unhappy reign. Speelman, commander of the VOC, drove rebel forces from the coast and established his base at Jepara, with company trading posts there and at Semarang, Surabaya, Rembang, Demak and Tegal. However, Amangkurat kept none of his agreements and it was not until the treaty of 1705 between Paku Buwono I and the VOC that Semarang was confirmed at coming under VOC control. The VOC finally moved its headquarters from the north-east coast there in 1708.

Semarang was an important port, but, like Jepara before it, its increasingly silted coast gradually diminished its importance. Paku Buwono II ceded Surabaya to the VOC in 1743 and the better port facilities gradually led Surabaya, particularly in the nineteenth century, to expand at the expense of Semarang.

There has always been a large and influential Chinese community in Semarang, though it has not always been appreciated. As a sequel to the massacre of some 10,000 Chinese in Batavia in 1740, Chinese forces attacked the Dutch in Semarang in 1741, with Paku Buwono II joining on the side of the Chinese. The Dutch were saved by Cakraningrat IV of Madura who relieved the city. The VOC drove off the attackers and massacred all the Chinese in the area.

Semarang was important in leftist politics in the early twentieth century with the Sarekat Islam and Communist Party developing strong bases there. In 1945, two months after the Japanese surrender at the end of the Second World War, the Japanese officers attempted to regain their authority from the Republican forces before the arrival of Allied troops and there was a bloody encounter between the two. Semarang also suffered in 1947 a Dutch attack on Republican forces.

The Kelenteng Sam Poo Kong temple is said to date from 1772 and is one of the oldest Chinese temples in Java. It is to be found somewhat lost in the large and chaotic Chinese quarter that retains its traditional habits and interest. The temple has some very good wood carving, particularly the beams, and a wealth of ritual objects and brassware.

There is another Chinese temple a little to the south-west of the city, known as Gedung Batu. This large cave houses the spirit of a Chinese admiral of the Ming dynasty who landed at Semarang in 1424.

The oldest church in Central Java, and the second oldest in Java, is also to be found in Semarang, the Gereja Blenduk (sometimes simply known as the Gereja Belanda or Dutch Church), in Jalan Jen Suprapto, south of the railway station. It was built in 1753 and is in the shape of a Greek cross, with a green copper dome raised on a high drum over the crossing. It has a handsome classical portico and inside a splendid Baroque organ surrounded by extravagant gilded cherubs. The church is in a rather dilapidated condition, somewhat symptomatic of the whole city, but is still imposing.

Semarang remains an important fishing port (all the fish on sale at Yogya and Surakarta come from here) and keeps a small naval establishment. It is essentially a commercial city and since it was taken over by the Dutch has always been so. It has a number of old warehouses and buildings with wrought-iron staircases on the outside, mostly in rather a poor state of preservation. The huge main square, Simpang Lima, with its massive government buildings, shows Dutch colonial planning on a grand scale. A large modern mosque, the Mesjid Baiturahan, is located on the north-west side.

Semarang has lively traditional theatre, especially the well-known Ngesti Pandawa troupe. It has a zoo, Tegalwareng, specializing in snakes. The Tugu Muda monument commemorates the five-day battle of Indonesian republican youth against Japanese troops at the end of the war and a special ceremony is held here every 14 October.

The new and elegant suburb of Candi to the south is perched on the hills overlooking the old city and is in pleasant contrast with the overall dilapidation of the central part. It is also, by virtue of its position, a good deal cooler.

This bustling city bursting at the seams is in total contrast to the gentle pace of Yogyakarta. It remains important in the administrative hierarchy, though has to give precedence to Yogya in culture, education and tradition.

# 8
# Practical Information

THIS chapter, which, if it were to include details of prices and places would be out of date before it were published, will be confined to generalities. For up-to-date details and prices, reliable travel agents should be consulted.

*How to get there*

Yogyakarta is about 500 kilometres from Jakarta and the most convenient way of getting there is by plane. There are several flights a day both to Jakarta and to Bali. The flight to both places takes just under an hour.

There are daily train connections from Jakarta and on to Surabaya; one train is air-conditioned. The train takes, of course, much longer but is cheaper than air travel. Long-distance trains arriving at Yogya tend to do so at unearthly hours of the night and are somewhat inconvenient.

There are many bus companies going to Jakarta, Surabaya, Denpasar and elsewhere. The fastest services are by night, the *bis malam*. Some of these are air-conditioned. The buses are quicker than the trains but are less comfortable. They are quite cheap.

*Where to stay*

Yogya has a great variety of accommodation, and new hotels and *wisma*, or guest-houses, are opening every year.

There are two main hotels in the main shopping street Malioboro, the Mutiara and the Garuda (the latter was an old-fashioned Dutch hotel and has been remodelled). The grandest modern hotel is currently the Ambarrukmo Palace, which is on Jalan Solo between the airport and the city; it has all the hallmarks of hotels of its kind, being expensive and impersonal, but it also incorporates three original buildings from the Kedaton Ambarrukmo when it was built as a secondary residence for Sultan Hamengku Buwono V. The buildings were remodelled at the beginning of this century and Sultan Hamengku Buwono VII spent his last year here. The buildings are in the form of pendopo of gracious proportions, somewhat dwarfed by the modern slab architecture by their side. The hotel is some way from the centre of town, like many of the newer hotels which have opened nearby.

There is a whole quarter south of the kraton, in the street known as Prawirotaman, given over to guest-houses, many with swimming-pools, all with attentive service and reasonable rates. These are recommended for those wishing for more character than the standard modern hotel. They are not particularly central and it is necessary to take becak from them to the main shopping streets. There are also several wisma off Jalan Solo not far from Gadjah Mada University; again they are not very central.

Around the station are several cheap hotels with rock-bottom comforts.

*How to get around*

There is no doubt that getting around Yogya is not particularly easy, for it spreads over large distances which are too great to be covered on foot. One obviously sees more when walking, but most often it is less effort to take a pedicab.

There is a kind of circular bus service going round the city, but the buses are erratic, overcrowded, dilapidated and uncomfortable. Furthermore, most tourists will have little idea where they are going. They are only recommended for hardy travellers. Watch your pockets if you take them.

If the weather is reasonable, one of the more exciting ways of getting around the Yogya area is on the back of a motor bike. Having a driver who knows his way around is also an advantage. The best way to find someone to do this is simply to ask. However, Solo, Borobudur and Dieng are too far away to be visited by motor bike.

For the still more hardy, motor bikes can be hired on making enquiries. It is wise to remember the practice if not the rule of the road in Indonesia, that the bigger vehicle always has precedence.

Hiring bicycles is easy and cheap; many places do this and the bicycle is the ideal means of getting around the centre of Yogya. If the bicycle is stolen, the replacement cost has to be paid.

Taxis are ancient and very difficult to obtain. They do not cruise the streets and have to be booked through hotels or guest-houses.

Minibuses go to most towns around and deliver one to one's destination.

Self-drive hire cars do not exist at present.

### *When to go*

The rainy season usually lasts six months, roughly from October to March, but this can vary, of course; sometimes the rains are early, sometimes late. It does not necessarily rain every day at this time, but usually does so in the afternoon, drizzling on into the early evening. Because of opening times, one is forced to do most of one's visiting in the mornings in any case, so the rainy season interferes only marginally with a visit to Yogya, except for the view of Merapi brooding over the city, which may be obscured at this period.

The dry season is not invariably dry either, but, given the choice, is the better time to visit Yogya. The Ramayana ballet at Prambanan is only performed in the dry season, usually from June to October.

### *What to wear*

Yogya can be cool at night, especially in the rainy season, and a sweater, pullover or even a jacket of some kind, especially if on a motor bike, will not be amiss at this time. For the most part, Yogya is a typical tropical city and one should dress appropriately. Sandals rather than shoes are common. Dresses, trousers and shirts should be washable natural fibres, preferably cotton. Formal dress for men is a long sleeve batik shirt, trousers, shoes and socks. Yogya is not very fancy in its style of dress; it is, however, conservative, and the outlandish or risqué will cause disapproval.

### *Where and what to eat and drink*

Yogya is not a gastronome's delight. *Gudeg* is a typical local dish, jackfruit cooked in coconut milk with eggs, soya bean cake and chicken mixed in. The only other specifically local dish is *mbok berek*, a whole fried chicken which is particularly tasty and eaten with *sambal*, hot chilli sauce. For the rest, the food available tends to confirm one's impression that the Javanese do not live to eat, though they do have a variety of snacks and titbits which are very tasty. There is a great tendency to put far too much sugar with everything and too little salt. Always ask for tea without sugar, *téh pahit*.

All the hotels and most guest-houses have restaurants of varying quality. Malioboro has a number of restaurants but if the truth were told none is exceptional. An open-air restaurant over some shops near the Garuda Hotel is popular.

*Rondé* is a special Javanese drink, red in colour, served warm and made of ginger, peanuts and slices of local fruit in syrup, and with balls of yellow glutinous rice. *Strop* is a syrupy iced drink with gelatinous sweets of gaudy hues. *Jamu* comes in various forms and is taken as a medicament for all kinds of complaints; it usually contains traditional medicinal herbs.

More conventional drinks are readily found and so are iced drinks like *es pisang*, banana ice, and *es nanas*, pineapple ice. Tip Top in Jalan Mangkubumi is a popular ice-cream parlour.

The fruits of Yogya are very good and very cheap; their variety is enormous. Hawkers up and down Malioboro sell them, especially near the main market, and almost any wayside stall has some local fruit for sale.

Imported food and drink, as well as a variety of local cakes and appetizers, can be obtained at Mirota in Malioboro.

Warung are small wayside stalls selling simple food and drink, often specially set up at night in the streets. They tend to specialize in *soto*, a spicy soup with bean sprouts and chicken or mutton, *bakso*, a noodle soup with meat balls, or *satay*, barbecued meat with peanut sauce.

Usually there are no health problems associated with food or drink in Yogya. However, hepatitis is not unknown and stomach upsets can occur. Western medicines are available without prescription in most *toko obat*, medicine shops, but *apotik*, pharmacies, usually require prescriptions.

For more serious problems there are many private doctors in the city and several hospitals.

*What to buy*

Yogyakarta, being the artistic centre of Indonesia and rivalled only by Bali for its arts and craft products, has enormous variety in souvenirs.

Batik usually comes highest on most people's shopping lists and is on sale everywhere. It can be bought in standard lengths (one by two and a half metres), or made up into shirts, dresses, table linen, etc. Price goes by quality; a first-class quality batik length can be very expensive indeed, but on the whole prices are low, given the workmanship that goes into the individually-produced pieces. The variety in design and colours, from very traditional to very modern, is enormous and there is something for all tastes.

Batik paintings are very popular and on sale all over town, with some of the cheapest (though not the most original) available in Taman Sari. Since they fold up to the size of a handkerchief and can be framed elsewhere, they are ideal for the air traveller limited by weight.

Leatherware is also popular with visitors. Bags, suitcases, sandals and belts are very cheap though the stitching and buckles are not always long-lasting. Wayang kulit figures are very popular. Old ones are difficult to find now and when available are expensive.

Wayang golek puppets are popular as decorative items and so are topeng masks, invariably new, since the masked dance is hardly ever performed now.

Silverware is the speciality of Kota Gede and the work is extremely elaborate. The local goldsmiths' work does not match that of the silversmiths.

The lightly-fired pottery of Kasongan is amusing, decorative and cheap, but tends to break on long journeys.

Antiques and curios these days tend to be mixed up; as the former become more rare, the latter dominate. Yogya is not particularly noted for antiques, though good buys are sometimes possible.

*Lurik* is a local cloth, made of cotton, and stronger than batik.

The surjan, or traditional man's shirt, is made of this material. It is attractive and cheap, but not light.

Gamelan instruments are available but not much in demand, given their weight and cost.

There is a somewhat deserted government crafts centre, the Desa Kerajinan or Crafts Village, on Jalan Solo between the town and the airport which has a selection of local crafts and batik.

# Glossary of Indonesian and Javanese Terms

| | |
|---|---|
| *abangan* | a follower of folk religion |
| *adu* | fight, contest |
| *alun-alun* | broad grassy area in the front and rear of a palace (alun-alun lor, the northern square, alun-alun kidul, the southern square) |
| *alus* | refined |
| *andong* | horse-drawn garry |
| *apotik* | pharmacy |
| *asrama* | student hostel |
| *babad* | history or chronicle |
| *bakso* | noodle soup with meat balls |
| *balai* | hall |
| *bangsal* | assembly hall |
| *batik* | cloth dyed after a traditional waxing process |
| *batin* | internal feeling, inner life |
| *becak* | pedicab |
| *bedoyo* | a court dance performed by girls |
| *benteng* | a fort |
| *bis malam* | night bus |
| *blangkon* | a male head-dress sewn into shape |
| *bonang* | a rack of bronze bell-like gongs |
| *bong supit* | traditional circumciser |
| *bupati* | regent |
| *candi* | ancient temple, shrine |
| *canting* | a small metal tracer used in making batik designs |
| *cap* | a metal printing block used in the batik process |
| *celempung* | a zither |
| *cempala* | wooden mallet used for sound effects and signals in the shadow play |
| *cempurit* | horn sticks of the shadow play figures |
| *cina* | China, Chinese |
| *daerah istimewa* | special district |
| *dalang* | puppeteer |
| *dalem* | inner quarters |
| *danyang* | guardian spirits |

| | |
|---|---|
| *demit* | place spirits |
| *dharmachakra* | (Sanskrit) the position of the Buddha teaching the Law |
| *dukun* | sorcerer, faith healer |
| *es nanas* | pineapple ice drink |
| *es pisang* | banana ice drink |
| *gamelan* | ensemble of classical Javanese musical instruments, the music of classical Java |
| *gedung* | building |
| *gender* | metallophone |
| *gendruwo* | spirits who play tricks |
| *gereja* | church |
| *gerobak* | ox cart |
| *gudeg* | jackfruit cooked in coconut milk |
| *gunung* | mountain, hill |
| *gunungan* | mountain-shaped shadow play figure symbolizing great natural forces |
| *jalan* | street |
| *jamu* | medicinal herbal drink |
| *jatilan* | a dance performed while riding a woven bamboo horse |
| *kasar* | vulgar |
| *kayon* | marker of the beginning and end of shadow play performances |
| *kebatinan* | meditation, mysticism |
| *kebaya* | blouse |
| *kendang* | drum |
| *keprak* | slit wooden box used as a sounding board by the puppeteer |
| *keputrian* | residence of the king's wives |
| *ketoprak* | modern popular theatre |
| *kidang* | deer |
| *komplet* | complete (gamelan komplet, a full gamelan of both tuning systems) |
| *kota baru* | new town |
| *krapyak* | fenced-in game reserve |
| *kraton* | palace |
| *kris* | ceremonial sword |
| *kromo biasa* | middle level Javanese language |
| *kromo inggil* | high Javanese language |
| *kromo madya* | lower middle level Javanese language |
| *kuning* | yellow |

| | |
|---|---|
| *lair* | external human behaviour |
| *lelembut* | possessing spirits |
| *ludruk* | popular East Javanese theatre |
| *lurik* | local cotton cloth |
| *Mahabarata* | cycle of stories of the warring Pendawa and Kurawa families |
| *makara* | mythical beast used as an architectural decoration |
| *manis* | sweet |
| *masjid ageng* | grand mosque |
| *mbok berek* | whole fried chicken |
| *memedi* | frightening spirits |
| *ngoko* | low Javanese language |
| *ngoko alus/sae*<br>*ngoko madya* | priyayi variants of low Javanese used when addressing persons of lower rank |
| *pangeran* | prince |
| *Panji* | cycle of stories of the Majapahit period about Prince Panji |
| *parang rusak* | broken kris, a batik design formerly reserved for royalty |
| *pasar* | market |
| *patih* | grand vizier, king's chief counsellor |
| *pelog* | tuning system of seven unequal distances to the octave |
| *pendopo* | open Javanese pavilion |
| *peranakan* | assimilated Chinese |
| *plengkung* | arched gateway |
| *prajurit* | soldiers |
| *pringgitan* | walkway, covered space |
| *priyayi* | aristocrat, upper class |
| *pulo* | island |
| *pusaka* | royal heirlooms |
| *raden* | title for male royal issue, prince |
| *raksasa* | giants |
| *Ramayana* | cycle of stories relating to Rama and his queen Sita |
| *ratu* | king, queen, monarch |
| *rebab* | two-stringed bowed instrument |
| *rondé* | warm drink of ginger, peanuts and fruit |
| *rukun* | harmony |
| *rumah makan* | restaurant, dining-room |
| *sambal* | hot chilli sauce |
| *sampur* | sash used in dancing |

| | |
|---|---|
| *santri* | orthodox Muslim |
| *sari* | fragrant |
| *satay* | barbecued meat with peanut sauce |
| *sawo* | a tasty brown soft tropical fruit |
| *selamatan* | ritual meal |
| *sendratari* | modern dance form, similar to ballet |
| *Serat Menak* | a cycle of stories about the Arabian prince Amir Hamzah |
| *serimpi* | a court dance performed by young girls |
| *sewu* | one thousand |
| *sinden* | singers accompanying the gamelan orchestra |
| *slendang* | scarf draped across the breast |
| *slendro* | tuning system of five almost equal distances to the octave |
| *soto* | spicy soup with bean sprouts and meat |
| *strop* | syrupy iced drink |
| *suling* | flute |
| *sultan* | monarch (of Yogyakarta) |
| *sumur* | well |
| *sundel bolong* | spirit of a prostitute with a hole in her back |
| *surjan* | traditional male jacket |
| *susuhunan* | ruler, emperor |
| *taman* | garden |
| *téh pahit* | bitter tea (that is, without sugar) |
| *tikar* | rush mat |
| *toko* | shop, store |
| *toko obat* | medicine shop |
| *tugu* | monument, obelisk |
| *tuyul* | familiar spirits |
| *upacara dalem* | regalia |
| *warung* | street stall selling food |
| *wayang* | theatre (usually the shadow play) |
| *wayang beber* | drama narrated as paper-scroll pictures are shown |
| *wayang golek* | theatre of three-dimensional stick puppets |
| *wayang klitik* | theatre of flat wocden puppets |
| *wayang kulit* | shadow play theatre with flat leather figures |
| *wayang topeng* | masked dance theatre |
| *wayang wong* | dance-drama performed by humans without masks |
| *wisma* | guest-house |
| *wong kecil* | 'the little people', ordinary people |
| *wringin* | banyan tree |

# Select Bibliography


Archer, Mildred and John Bastin, *The Raffles Drawings*, Kuala Lumpur, Oxford University Press, 1978

Brandon, James R., *On Thrones of Gold*, Cambridge, Mass., Harvard University Press, 1970

_____, *Theatre in Southeast Asia*, Cambridge, Mass., Harvard University Press, 1967

Dalton, Bill, *Indonesia Handbook*, Rutland, Vt., Moon Publications, 1978

Dumarçay, Jacques, *Borobudur*, Kuala Lumpur, Oxford University Press, 1978

_____, *The Temples of Java*, Singapore, Oxford University Press, 1986

_____, 'Le Taman Sari (Étude achitecturale)', *BEFEO*, Vol. LXV, 1978

_____, 'Notes d'architecture javanaise et khmère', *BEFEO*, Vol. XXI, 1982

_____, *Candi Sewu et l'art bouddhique du centre de Java*, Mémoires Archéologiques XIV, Paris, École Française d'Extrême Orient, 1981

Geertz, Clifford, *The Religion of Java*, Chicago, University of Chicago Press, 1960

_____, *Peddlers and Princes*, Chicago, University of Chicago Press, 1963

Geertz, Hildred, *The Javanese Family*, New York, Free Press, 1961

Hall, D. G. E., *A History of Southeast Asia*, 3rd ed., Macmillan, London, 1968

Holt, Claire, *Art in Indonesia, Change and Continuity*, Ithaca, NY, Cornell University Press, 1967

Jessup, Helen, 'Sonobudoyo Museum, Yogyakarta', *Orientations*, Vol. 13, No. 11, November 1982

Kartodirdjo, Sartono, *The Pedicab in Yogyakarta*, Yogyakarta, Gadjah Mada University Press, 1981

Kunst, Jaap, *Music in Java: Its History, Its Theory, and Its Technique*, 2 vols, 2nd ed., The Hague, Martinus Nijhoff, 1949

Lindsay, Jennifer, *Javanese Gamelan*, Kuala Lumpur, Oxford University Press, 1979

Mulder, Niels, *Mysticism and Everyday Life in Contempoary Java*, Singapore, Singapore University Press, 1978

Raffles, Thomas Stamford, *The History of Java*, London, 1817 (Kuala Lumpur, Oxford University Press, 1978)

Rendra, *Ballads and Blues: Poems*, Kuala Lumpur, Oxford University Press, 1974

Ricklefs, M. C., *A History of Modern Indonesia*, Macmillan, London, 1981

______, *Jogjakarta under Sultan Mangkubumi, 1749–1792*, London, Oxford University Press, 1974

Scidmore, Eliza R., *Java, Garden of the East*, New York, 1899 (Singapore, Oxford University Press, 1984)

Soemargono, Farida, *Le Groupe de Yogya, 1945–60*, Paris, Cahiers d'Archipel No. 9, 1979

Sutherland, Heather, *The Making of a Bureaucratic Elite: The Colonial Transformation of the Javanese Priyayi*, Singapore, Heinemann Educational, 1979

Van Ness, Edward C. and Shita Prawirohardjo, *Javanese Wayang Kulit*, Kuala Lumpur, Oxford University Press, 1980

Williams, Glen, *Guide to Yogyakarta*, 2nd ed., Yogyakarta, Kanisius, 1975

Wurtzburg, C. E., *Raffles of the Eastern Isles*, London, 1954 (Singapore, Oxford University Press, 1984)